Climbing the Limitless Ladder

A Life in Chemistry

IISc Press-WSPC Publication

Climbing the Limitless Ladder

A Life in Chemistry

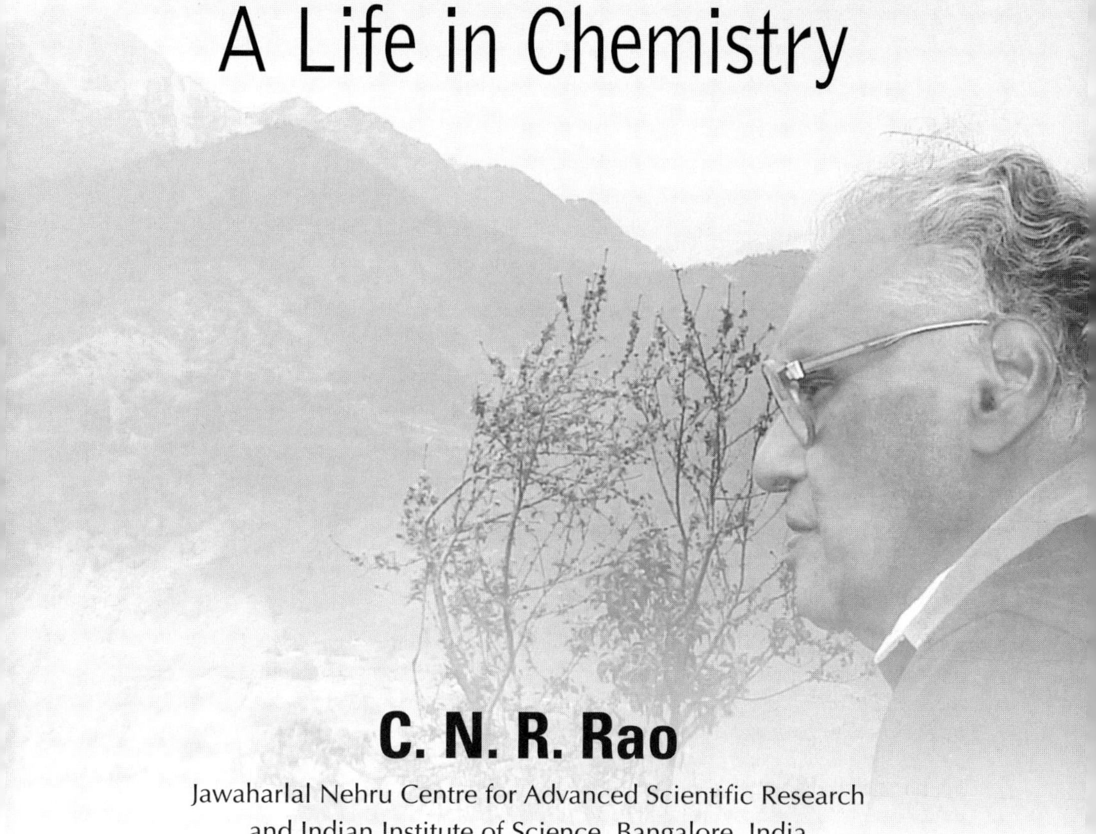

C. N. R. Rao

Jawaharlal Nehru Centre for Advanced Scientific Research
and Indian Institute of Science, Bangalore, India

IISc Press

World Scientific

NEW JERSEY • LONDON • SINGAPORE • BEIJING • SHANGHAI • HONG KONG • TAIPEI • CHENNAI

Published in India by

Cambridge University Press India Pvt. Ltd.
under the imprint of Foundation Books
Cambridge House
4381/4 Ansari Road
Daryaganj,
New Delhi - 110 002

This reprint edition 2011

ISBN- 13: 978-81-7596-912-4

This edition of *C. N. Rao / Climbing the Limitless Ladder : A Life in Chemistry,* is published by arrangement with World Scientific Publishing Co. Pte. Ltd.

This edition is for sale in India, Pakistan, Sri Lanka, Bhutan, Nepal, Maldives, Afghanistan and Bangladesh only, not for export elsewhere.

Published by Manas Saikia for Cambridge University Press India Pvt. Ltd. and printed in India at Sanat Printers, Kundli, Haryana.

Dedicated to

Indu

Prologue

For more than a decade, many of my friends in India and abroad have been asking me to write about my life in science. The first time I was asked this, I had just turned 60. I felt shy writing about myself since I was not sure that I was eminent or old enough to write an autobiographical memoir. The only difference since then is that I have become older. I am now 75 years old and have spent more than half a century doing research in India. Many of my colleagues and friends have convinced me in the last few months to write about my career in science. Their view is that my experiences in doing science for several decades may inspire young people.

I have written this book in the form of an autobiographical essay. It was difficult to subject myself to a self-examination. I had to use a telescope as well as a microscope. More importantly, I had to be honest and truthful. I have avoided exaggeration and the use of superlatives as far as possible and written this book in simple language, recognizing that simplicity of expression demands clarity of thought. I did not want the book to be a record of events or a progress report. Neither did I want it to be a directory listing all the people that I have met. I have written it in a way that hopefully conveys a message. I have included a few of my essays as appendices to elaborate on some of my beliefs and aspirations mentioned in the text.

In the last few decades, there have been considerable changes. India has changed, and the world has changed. Science has changed, and the way we do science has changed. I have tried to reflect on these changes and also the way that I have tried to accomplish something in science, working in India. I do hope that the book will provide an insight to a feeling of how, in spite of many limitations, one can try to *climb the limitless ladder of excellence*. I would like to be excused for any omissions and errors in judgment.

It is with great pleasure that I dedicate this book to Indu who has stood by me for over half a century.

I am delighted that the book is being published by World Scientific in collaboration with the Indian Institute of Science. I am proud of my long association with the Institute which has been the birth place of many fine scientists and great institutions.

C.N.R. RAO

May 2010
Bangalore, India

Contents

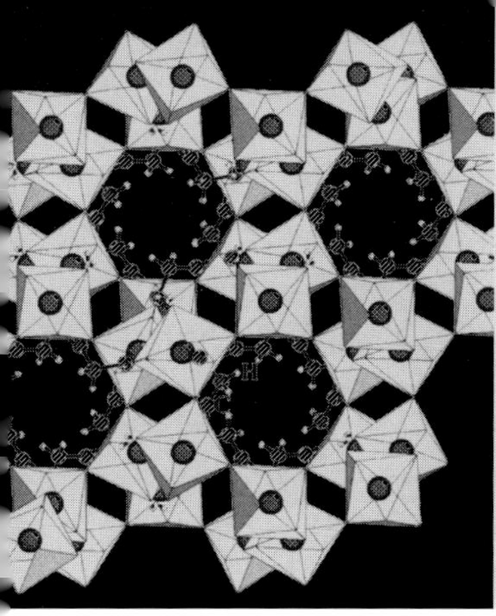

1

Those Days

In the 1930s and 40s, Bangalore was a quiet and peaceful city, with rustic surroundings and a leisurely atmosphere. Not too far from Bangalore, there were tigers in the jungles. In the south of Bangalore, elephants roamed freely. Life in the city was simple and there was little room for excitement. One ate simple food at home at fixed hours and went to school. Young people played simple games in the evening, and only a small number of them played cricket and other modern sports. Roaming through the streets in the vicinity of one's residence or spending time with friends was the main entertainment. Roads in Bangalore were empty except for an occasional bus of the Bangalore Transport Corporation. It seemed everyone walked to school or college and a fortunate few owned bicycles. The mode of transportation for most people was walking. There were horse-drawn carts to go to the railway station if the luggage was limited. Bullock carts were available for transferring heavy goods. On any occasion, if one wanted to shop in the city market or in the big shopping area in the centre of the city, all he/she had to do was to walk for a few miles. One could also take a bus to the city market.

The simple life in Bangalore among adults was made interesting by frequent visits to restaurants for coffee, two sharing a cup (one by two) being common. Alas, children had no chance to visit these cafes since money was scarce. Vidyarthi Bhavan and MTR were favourite spots for dosa and other delicacies. Bangaloreans loved life, and restaurant-visits were common. There were famous little shops that specialized in selling items like spiced peanuts which cost only a few paisa. Nani's shop in Gandhi bazaar was one of them.

Weather in Bangalore was always nice. It was considered an air-conditioned city, except for a few hot days when we would occasionally see a person holding an umbrella. One did not seem to feel the heat of summer those days. There were no fans in houses or class rooms. One had not heard of air conditioners. Bangalore was green and it was no surprise that the fine weather of Bangalore and its serene atmosphere created the reputation of the city being a pensioners' paradise.

Bangalore was a major city of the erstwhile Mysore state, ruled by the Maharaja of Mysore. The state was progressive. Mysore state was one of the few states with steel mills, paper mills and dams for irrigation. The Maharaja lived in Mysore city, even if all the main offices were in Bangalore. Mysore seemed like a beautiful toy town. A special occasion there during the Dasara celebrations was the procession of the Maharaja riding the palace elephant on the last day (Vijayadashami) of the season. The day in Hindu culture signifies victory of good over evil. The palace is lit up by 100,000 bulbs on that day, making it appear like a palace in a fairy tale. The Maharajas were good to the citizens, Krishnaraja Wodeyar being the best example. The state had the benefit of a representative assembly to which people elected members. Even the council of ministers had some members from the public. Dewans of Mysore were outstanding planners and administrators. For example, the beauty of Bangalore owed much to Dewan Sir Mirza Ismail. Dewan Sir Seshadri Iyer was responsible for allocating the land and other facilities to establish the Indian Institute of Science in Bangalore.

Mysore state had eight districts, and the hills and plantations of the *malnad* districts of the state were most attractive to visit. Malaria was rampant in the *malnad* area, until it was eventually eradicated in the 1950s. There was at least one government high school in each district of

Mysore state, but there were only two colleges, one for science in Bangalore (Central College) and the other for humanities in Mysore (Maharaja's College).

Electricity was not common in most parts of India, including in Mysore state. One still remembers the advent of electric bulbs in the homes of Bangalore. Till the late 1940s, human refuse was manually cleared by fellow human beings which was terrible to bear. Flushing toilets got introduced in the city in the late 1940s. Very few people had radios in their homes. Telephones were unknown and communication with another person required a visit or writing a letter. The postman was an important person in homes and offices and people eagerly awaited the arrival of postmen. The absence of modern communications facilities (which are considered to be essential today) did not seem to take away much from the happy life that citizens of Mysore state enjoyed at the time. Kannada was the language of the state. Bangaloreans, however, merrily mixed English with Kannada words.

It seemed Bangalore comprised one long road connecting Basavanagudi in the south and Malleswaram in the north. In between, were the city market and some shopping streets, and areas like Chamarajapet and Seshadripuram. After Malleswaram 18th cross in the north, there was the Indian Institute of Science, and one hardly saw people in this area, specially after working hours. There were two fine gardens in the city, Lalbagh and Cubbon Park, which were not crowded and provided much enjoyment to Bangaloreans. Lalbagh had unusual trees from everywhere and was a walkers' paradise. Cubbon Park, in the centre of the city, had a bandstand and also a nice restaurant. To the east of Bangalore, was the cantonment area visited by very few Bangaloreans. Entry to some of the Cantonment areas was forbidden to Indians during the British days. People of Basavanagudi always considered themselves as the custodians of the culture of Bangalore.

Money was hard to come by those days. People were thrifty and one rarely saw exhibition of wealth. Most people in Bangalore came from middle class or poor families. The few who were rich were not noticeable. There were no politicians crowding the city at that time. There were no billboards and larger-than-life cut-outs of politicians and other lesser known persons. Many who became politicians later were freedom fighters or nationalists at that time, and most of them wore simple clothes (Khadi, commonly) and some of them preferred to walk bare-footed. Occasionally, there would be some excitement in Bangalore in connection with the visit of a VIP, or due to a seasonal festival. *Karaga* elicited an enthusiastic response from a section of the people. [Karaga is a festival to celebrate the goddess of Shakti — the Priest of the Dharmaraya temple decorated in feminine attire usually walks with a delicately balanced earthen pot on his head. The procession that accompanies him has men demonstrating their wizardry with swords. The earthen pot represents the goddess of energy and power. It is celebrated even today.] Ganesh chaturthi and Rama Navami were other socially important festivals. Bangalore looked forward to the month-long music concerts held all over the city during Rama Navami celebrations. I remember the visit of the Viceroy of India in the early 1940s when all the school children were lined up along the streets.

The most significant events during 1940–47 were the nonviolent processions and demonstrations against British occupation and the cry of people for freedom. The freedom movement was specially strong in 1942 and continued till 1947 when India eventually got freedom from the British. The freedom movement in Bangalore did not seem to be as intense as in northern India, and yet there was occasional shooting and lathi charge in the streets of Bangalore. There were rumours of some of the leaders of the movement going into hiding. One of the experiences that I cannot forget relates to the conditions in Bangalore and elsewhere during the Second World War and a few years later. Food and cloth were rationed. Rice, wheat, kerosene and cloth could only be obtained through ration cards.

In Bangalore of the above description, I was born on June 30, 1934 in Basavanagudi, in the home of my maternal grandparents, located not far from Bugle Rock (close to the bull temple) and the famous temples of Ganesha and Hanuman. I do not remember the days that I spent in that location. I remember, however, that my grandparents moved to another house close by on Nagasandra road (now D.V. Gundappa Road). Close to our house was Mr. D.V. Gundappa, the famous literary personality. Mr. Nittoor Srinivasa Rao (later Justice and the first Vigilance Commissioner of India) was frequently seen in the locality, taking walks with Mr. Gundappa. Nittoor lived upto the age of 100 and seemed to know everyone in microscopic detail. Other companions of Mr. Gundappa during his evening walks were Mr. R.L. Narasimhiah (a fine physics teacher in Central College) and Mr. Somashekhara Rao (a lawyer). I mostly associate my childhood with the corner house on Nagasandra road with two champak trees in front. At the end of the road was the house of Nettakallappa who took pride in hunting tigers. His house was full of stuffed tigers and other animals. Masti Venkatesha Iyengar (Masti), a renowned literary personality, lived not far from us, in Gavipuram.

My maternal grandfather (Belur Seshacharya) was a teacher of Kannada and Sanskrit and taught at the Ladies Intermediate College and later in some of the schools. He told me once about his childhood. His father (my maternal great grandfather) apparently took sanyas (became an ascetic), and left home when my grandfather was very young. My grandfather was brought up by his mother. One fine day, when my grandfather was grown-up and well established, my great grandfather showed up in a sanyasin's garb and breathed his last. My grandfather was a strict disciplinarian and dressed in an orthodox fashion, with a brocaded turban, close-collar coat and dhoti. I do not remember him taking a bus or any other vehicle to go anywhere in Bangalore. He always walked. My maternal grandmother (Padmavati Bai) was an extraordinary person, full of life, and she dominated the home. She

Rao's maternal Grandmother, Padmavathi Bai

could compose instant poems and songs, and sing as well. She was a great conversationalist and social being. She was an excellent cook and her filter coffee was famous amongst friends and relatives. For a person with two or three years of schooling, she commanded an extraordinary presence. She would enthrall her listeners with her stories of people and places. I remember an occasion when Rajaji (C. Rajagopalachari, the first Governor-General of India) visited Mr. Navaratna Rama Rao who lived close to us. My grandmother took good strong coffee for Rajaji and sang a few Purandaradasa's compositions (devaranamas) for him. My grandfather and all of us were dumb-struck by her boldness and natural charm. She had a great sense of humour which compensated for the very silent, serious nature of my grandfather. She was generous to a fault and there was no limit to visitors, free loaders and all kinds of relatives who came for lunch or for short stays.

My maternal grandparents had three sons and four daughters, and my mother was the eldest of the children. My mother (Nagamma, born 1908) was a very orthodox person and prayed for several hours in the day, starting at five in the morning. She had little use for worldly possessions and found complete happiness in her prayers. She was self-educated and was much interested in current affairs, considering that very few women went to high school those days. She was an ardent reader of newspapers. My father (H. Nagesa Rao, born 1905) worked in the education department of Mysore state. He was an M.A. in History, Economics and Politics and had degrees in Education. He was highly

Rao's Mother, Nagamma at her daily prayers

Rao's Father, Nagesa Rao

disciplined and was a stickler for punctuality. He had to have coffee at 6.00 a.m. in the morning and tea at 2.15 p.m. His main hobby was reading newspapers and books. In fact, whenever I think of him, the image that comes to my mind is of my father holding a book in his hand. He took extraordinary trouble about the food we ate, choosing the fruits and vegetables himself. He did not believe in astrology and was a great champion of family planning. Despite the contrast in their personalities, my parents shared respect for scholarship and education. They were both progressive in their outlook. My father was very strict with financial matters as well. Yet, when Vinobha Bhave started the bhoodan movement, he gave away our ancestral lands near Chintamani to the movement, to be distributed to the toiling farmers.

My father came from a family of landlords not too far from Bangalore on the border of Andhra Pradesh. The lands had been granted by the Maratha Government under Shivaji who ruled Bangalore, Kolar and nearby places at one time. My forefathers who collected taxes for the Maratha Government had the title Deshmukh. They had hailed from Kolhapur area several decades earlier. Apparently, my paternal great grandfather (Deshmukh Ananda Rao) was well off, but was a spendthrift. My paternal grandfather (Deshmukh Hanumantha Rao) tried hard to bring stability to his family by regaining some property. He lost his wife quite early and had to raise a family of two sons and three daughters with much difficulty. My father was the eldest son and the first one to go to college in the family. He had to manage his studies with the meager funds available to him. My contact with my paternal aunts was limited since they all lived away from Bangalore in Kolar district, but my paternal uncle (Narayana Murthy) used to be a constant visitor to our house. An interesting feature of all my relatives on my father's side was that they were all adept in classical music. Some were professional musicians as well. I remember the day in 1939 when my grandfather died in his sleep. I did not get to know him well or live with him for long. I was fond of one of my paternal aunts (Puttamma).

Unfortunately, her husband (a music vidwan) disappeared when the family had gone for a picnic on a river bank. My aunt did not recover from the shock.

I was very close to my maternal aunts and uncles. I must specially mention my aunt Susheela, who lived in the north of Bangalore, whom I visited often. She was a lovable person and a fine cook. My uncle and she were always very good to me. Another aunt, Kamala, lived far away in Vadodara and there was no way that I could visit her. Another aunt of mine, Indira, was a fine person who died prematurely. Her husband had put her in a horse-cart and sent her to my grandfather's place when she was close to having a baby. She died in front of my grandfather's house, as she entered it. One of my maternal uncles (S.R. Swamy) was an expert in Montessori education and was dedicated to the cause of nursery education through out his life. Another maternal uncle of mine (Murthy) was a little older than me and was my classmate through school and college. He was an amateur artist and we used to play together. The third maternal uncle of mine (Sridhar) was much younger to me.

I must mention an important aspect of my background. Families of both my mother and father were staunch believers in Madhwa philosophy. Madhwa ideology provides a practical way to lead one's life. According to Madhwacharya, there are two distinct worlds — there is the world of adhyathma (the spiritual world), but equally important is the world we live in. In this real world, we are encouraged to live properly and serve mankind. There is much sense in this simple statement.

I did not go to primary school and studied at home under the guidance of my mother. She was an extraordinary teacher. Every day she would make me do arithmetic and read lessons. She was terrific in doing mental maths and could multiply and divide large numbers so easily. She would tell me stories, particularly from Mahabharatha, Ramayana and Bhagavatha. By the age of 7 or 8, I knew probably most of Ramayana and Mahabharatha. I had to narrate them back to her on some days. I did not enjoy sports much, and my main hobby was talking to people in the neighbourhood. I started going to middle school at the age of six. (There was no age restriction for admission to schools those days). My classmates were all older than myself, but this did not affect me in any way. I used to tutor many of them in English and Maths. I spent time with my friends near my home in the evenings, mostly talking. I completed the lower secondary (LS) examination (equivalent to 7th class/standard now) of the Mysore State in 1944 with a first class. The examination was held state-wide with great formality and external examiners evaluated the answer scripts. Passing the LS examination with distinction was considered to be a great accomplishment at that time, but it made no impression on me. I was 10 years old. My father gave 4 annas (25 paise) as a reward to eat an ice cream in MTR. The biggest reward that I received was one rupee from my grandmother's aunt (Sundarakka). An important news that I remember when I was a student in the middle school was the death of Rabindranath Tagore (1941). I recollect my mother telling me about the greatness of Tagore, and how he had received the Nobel Prize for literature.

When I went to high school, I could study either in Kannada medium or in English medium. My father insisted that I should study in Kannada medium since the subjects were best learnt in the mother tongue. I became a Kannada medium student, but at home my father spoke to me in English to make sure that I did not suffer from the lack of knowledge of English. I went to more than one school during my high school education since my father was being transferred to different locations in the education department of the state. This enabled me to see places in Chickmagalur and Shimoga districts, the beautiful malnad areas of the state. I used to accompany my father to interior areas, specially hills and jungles, encountering wild animals in their natural habitat. This meant walking for miles with my father and some attendants. We saw elephants, bears and tigers in the wild on some occasions. I enjoyed my time in high school, particularly my science classes. I had outstanding science teachers in high school and I specially remember my chemistry teachers, Mr. Shivarudrappa, Mr. P.S. Narayana Rao and Mr. S. Krishnamurthy. They were inspiring and showed excellent experimental demonstrations in school which I remember to this day. When in high school, I participated in debates and essay writing (both in Kannada and English), and took part in Kannada plays. One of my interests during the school days was to read Kannada literature. This was mainly because my mother enjoyed reading and we would read books together. I would bring many books for my mother and read them with her. I even tried my hand in writing poems in Kannada. When I was 11 years old, I took a little poem to Masti and asked him if he could publish it in a journal edited by him. Masti was very kind and told me to come back later with a longer poem.

One occasion that stands out in my memory is the visit of Professor C.V. Raman to my school, Acharya Pathashala, in Basavanagudi. Roddam Narasimha who was my schoolmate also remembers the occasion. Professor Raman gave a fantastic lecture and unveiled the photograph of Madame Curie. This made an extraordinary impression on me. It is possible that it was at that moment I took the subconscious decision of becoming a scientist. Professor Raman asked our teacher to bring two or three of the good students from the school to visit his laboratory in the Indian Institute of Science. I was lucky to be one of them. Prof. Raman spent more than an hour telling us about what he was doing in the laboratory.

I completed the Secondary School Leaving Certificate (SSLC) examination of the Mysore State in 1947 with a first class. SSLC had two categories ECPS standing for eligible for college and public service and EPS meaning eligible for public service. Those with EPS could not go to college for further studies. I wanted to get a Bachelor's degree in science. This took four years after S.S.L.C. I spent the early part of this period in Shimoga (Intermediate College) and later joined Central

CNR Rao as a college student (1947)

College, Bangalore. India obtained freedom in August 1947, when I had just entered college.

Mysore state did not readily join the Republic of India soon after India gained independence. The Maharaja had been wrongly advised by the Dewan and others. There was serious agitation in the state against the state government which lasted for a few months. It was exciting to participate in processions and slogan shouting. On a few occasions, I gave public speeches with some of the leaders. The college was closed during that period. Mysore state soon joined the Republic of India and we had a democratic government installed with leaders of the congress party ruling the state. A remarkable feature of the first Independence Day was the smooth transfer of power from the British to the Indians. Partition caused much pain and agony for the people in Punjab and Bengal, but we in Bangalore did not fully comprehend the enormity of this occurrence.

August 15, 1947 has a special place in my life, and even today if I close my eyes, I can hear the great speeches of that day. But soon after, Gandhiji was assassinated (January 30, 1948). I was stunned by the event. It took me a long time to recover from the terrible loss of Gandhiji whom I had loved. I had seen him once when I was in the second year of high school in Acharya Pathashala. Some of our school teachers took us to Madras (now Chennai) to see Gandhiji. I felt that it was most ironic that the sage of nonviolence was brutally killed by a mad, violent man.

As an undergraduate student in Central College, Bangalore, I studied physics, chemistry, mathematics and other subjects. We had good teachers in physics and mathematics. I had to depend on my own to learn chemistry. I must remember Prof. B.S. Madhava Rao (a collaborator of Homi Bhabha) for his fine mathematics classes. It seemed as though my best teachers were during my school education. During my college days too, I participated in debating, writing short stories (in Kannada) and in Kannada plays. I was secretary of the Sanskrit Association in the college. I do not remember anything academically interesting during the four years of college. I, however, still remember a debate that I attended with Norman Cousins and S. Radhakrishnan as speakers. The topic was related to the advantages of English as a language compared to Sanskrit. Cousins told the audience how one can express oneself in many ways and exactly in English. For example, he asked what the word was in Sanskrit for "Precision". Radhakrishnan was at his best and asked what the word in English was for "Dharma", and concluded that he would rather have dharma than precision.

One of the difficulties that I faced in college (and even earlier) was from studious classmates, who were always talking of the so-called important topics for the examinations, and were studying day and night. They would constantly ask me whether I had studied one topic or the other. I could only read something once or twice and could not read the same thing over and over again. I had to keep mum when my good friends would argue on what the important topics were for the exams. In retrospect, I understand the difficulty with exams that we faced at that time. To get 60% marks in an exam was considered remarkable. Teachers would say that the answer to a question was perfect and give 6 or 7 marks out of 10. English exams were especially bad. I fancied myself as being adept in English, but I never got good marks in the exams. However, in spite of everything, I completed my B.Sc degree examination with a first class. I was one of the very few students who got first class in the B.Sc examination of 1951 of the University of

Mysore. My father was pleasantly surprised since he always felt that obtaining a first class in B.Sc was next to impossible. After the results were released, I rushed to my aunt Susheela to tell her, and then to a temple as per my mother's instructions.

My undergraduate days were marked by activities which made me nationalistic. I became a great admirer of Pandit Nehru and a few other leaders. I participated in activities which involved voluntary work and editing magazines. We played khokho, kabbadi and such games during evenings. I used to wear a Gandhi cap up to my second year in college and participated in meetings and lectures where national leaders and other personalities were involved. I regularly attended lectures at the Gokhale Institute of Public Affairs (run by Mr. D.V. Gundappa and Nittoor) and at the Indian Institute of World Culture. I got to know Dr. H. Narasimhiah of National College, Bangalore, during my undergraduate days because of my interest in Gandhian ideas and in voluntary work. (Narasimhiah became a popular educator in Bangalore and was highly respected in the community.) One of the occasions that I remember is a session of the congress socialist group with Yusuf Meharauli, Narendra Deo, Jaiprakash Narayan and others. We were involved as volunteers in making various arrangements. We used to recite certain parts of the Gita once a week.

I was not too sure of what I wished to pursue professionally during my undergraduate days. There was considerable temptation to take up engineering or medicine, after my Bachelor's degree in science. Some of my relatives suggested that I could go for civil service. Fortunately, my parents told me to do what I liked best. This helped me to arrive at a satisfactory decision on my own. I had a romantic notion about doing research as a scientist, but had nobody to guide me. When I was in the first year of the B.Sc programme, I had come across a journal published by the University of Mysore containing research papers. One of the papers on magnetic susceptibilities of some compounds had Mr. Venkataramiah (a physics lecturer in Central College) as an author. I went to him to enquire whether I could do research in summer, but he did not encourage me. I used to see people who worked at the Indian Institute of Science (IISc) as research scholars. Somehow, they did not seem so inspiring. The research students from the Institute did not tell me much except that getting a Master's degree by research could take an indefinite period. At that time, IISc did not give degrees and one had to get them through Bombay and other universities. One of my teachers in Central College (P. Mallikarjunappa), who himself was not a research scientist but had received a Master's degree in chemistry from Banaras Hindu University came to my rescue. He told me that for a Master's degree in that university, doing research was part of the programme and showed me a research paper that he had published in the Journal of Indian Chemical Society. He urged me to go to Banaras Hindu University to do a Master's degree. This made a big difference in my life. I would otherwise have joined IISc for a diploma in chemical engineering, to be followed by a research degree from a university. A Master's degree from Banaras which included research seemed to be the best option of that day. I wrote a letter to the head of the department at the university (Prof. S.S. Joshi), and received a telegram a few days later admitting me to the M.Sc programme in chemistry. I did not look for any other option.

I often dream of those days in Bangalore. I cannot forget the happy days with my maternal grandparents, and my grandmother's innumerable stories and anecdotes. I remember walks with my father, specially in the jungles. Even today, certain individuals, incidents and stories appear like flashes at the most unexpected times.

I think of Kuchela, my dearest friend during the four years of college who migrated to Canada (and is no more).

I can feel Kitti (who claimed to be a distant relative of mine) tapping on my shoulder in a music concert, just to tell me about the progress being made by his innumerable children. He would say, you know, my 7th one is much smarter than the 6th one (he never used their names).

Then he would say, do you have any money in your pocket? Let us have a one by two (coffee) somewhere. Kitti was a poor man, with a large family, but I did not know a happier man.

I recollect the story told by Navaratna Rama Rao who lived close to our home. When he was a judge, he scolded a thief accused of day-light robbery in a house. When he asked the thief, why he stole household goods, the thief got terribly angry and retorted, "don't accuse me of such petty thefts. I have always gone for major robberies worth something. I have some standing in my profession".

I think of Mr. Shastri, my grandfather's friend, who was always complaining about his wife and telling the entire town how he would have nothing to do with her. She, however, bore a child every year.

I think of the great playright and satirist, T.P. Kailasam, walking around in his underwear with a cigarette in his mouth and a cigarette tin in his hand.

Narasinga, the imbecile young man, who wandered the streets of Basavanagudi talking to anyone he came across, became a martyr in 1947. He was one of those few who got shot by the police in the independence movement.

I recollect little of the early days of the parliament of independent India. I must, however, recount something that I heard from a senior M.P from Karnataka (I do not know whether it is true). A member of parliament from Karnataka got up and started talking in Kannada. He had not spoken before. Everyone got confused. Pandit Nehru suggested that it be translated. On translation, it was found that the member had only said, "I am feeling hungry. It is getting late for lunch".

I often think of Dr. M. Shivaram and the harmless humour in the magazine Koravanji, that he edited.

I frequently dream of those dreaded final exams and wake up with a start. I dream that the exam was in physics, but I would have prepared for chemistry. I dream of the Sanskrit exam or of the English exam on Shakespeare. I appear for an exam on Macbeth, but the exam would be on Hamlet. The worst dream is of a maths exam. I do not recognize any equation or understand any question. I hate final exams of the kind that I took which once and for all decided what one did in life.

The dream that I cherish most is the one of the first independence day on August 15, 1947 when my neighbours and friends went around the streets with bells and gongs.

Where have they all gone,
those simple men with great character,
those unsung heroes with fine minds,
but little riches.
Alas, times have changed,
there are many making loud noises,
some with much riches, but few with
sense and substance.

– Anon

2

Pilgrimage for Knowledge: Banaras to Berkeley

My search for a direction was answered by admission to the Banaras Hindu University (BHU). I had decided to become a chemist and told my parents of my desire. As always, my parents wanted me to do what I liked most. They trusted my judgement completely. Studying in Banaras implied that my parents had to send me around Rs. 75 per month for expenses. This was a big amount those days, and my father said that he would somehow manage it. I had done much better in physics than in chemistry in my B.Sc and I hoped that someday I would study physics as well.

CNR Rao with his parents

On a day in June 1951, just before my 17th birthday, I left Bangalore by train in an unreserved third-class compartment to Banaras (we no longer have third-class railway compartments). I found on the way a few other Bangaloreans going to Banaras for studies. BHU was a national university those days with students from all over the country. I was

Rao in BHU campus

truly impressed by the beautiful BHU campus. One wonders, even today, how the great founder of BHU, Pandit Madan Mohan Malavia, could envisage and plan such a campus so long ago. Soon after I reached the University, I was allotted a hostel for accomodation. A few days later, the university started its session with a large gathering of students in the sports field where the great musician, Pandit Omkarnath Thakur, sang the *kulgeeth* (university anthem). I went to the department and met Prof. S.S. Joshi, the head of the department and opted for physical chemistry as my major field. Prof. S.S. Joshi was a D.Sc from London University and had worked in the University College with F. Donnan (student of Ostwald and Van't Hoff, the earliest physical chemists and Nobel Laureates in Chemistry). The M.Sc classes started right away. The work mainly involved carrying out experiments in the chemistry laboratories. My class had students from all over India (Assam, Andhra, Tamil Nadu, Kerala, Maharashtra, Bengal and Karnataka) and surprisingly there were only two or three from Uttar Pradesh. Coming from Bangalore (Mysore University), I suffered some disadvantage. My classmates were always bragging about the high standards of their colleges and universities, particularly in laboratory training. I had to

Chemistry Department — BHU

suppress my fear of not having been so well equipped in terms of chemical information.

A week after the course started, Prof. Joshi asked me to begin my research for the M.Sc degree, instead of waiting for a year. The general practice was that during the first year, the student took courses in the two minor subjects (organic and inorganic chemistry in my case). During the second year, he/she concentrated on the major subject (physical chemistry in my case) and did research on a given topic. It was a surprise to be told to begin research on electrical discharge in gases during the first year. My immediate adviser was Dr. B.R. Marathe who had just completed his Ph.D degree. The problem was to study the effect of light on electrical discharge in gases (popularly called Joshi effect). There was no alternating current (AC) electricity in Banaras at that time. There was only direct current (DC) electricity. One had to convert DC to AC to do research, and this was possible only at nights. All my research was, therefore, to be done during night and I became a regular night owl in the laboratory.

The first few days in Banaras were difficult since I was not used to the hot weather or the food that was available. Soon I could join a mess that offered food that was somewhat similar to that in Bangalore. (Gradually, I began to like North Indian food, specially samosas and paranthas.) Living conditions in the university hostel were satisfactory except for the toilets that had no flushing facilities. By paying one rupee a month, one could get people to clean the toilets before use. I disliked doing this, but there was no option. One had to bathe in cold water which I got used to. Throughout my stay in Banaras, winter or summer, I bathed in cold water. My expenses for food and other basic needs came to around 60 rupees a month.

We did not have many lectures by faculty members in BHU, but some of them made a great impression. In particular, the organic chemistry lecturer, Dr. R.H. Sahasrabudhey, was a unique person. He was self-effacing and allowed his research students to publish papers without his name. Prof. Joshi gave one lecture during my second year. He was a good lecturer. We had to give seminars, and my topics were cosmic rays and atomic hydrogen. I could see many well-known scientists in BHU, Lord Todd of Cambridge (Nobel laureate in chemistry) being one of them.

The research atmosphere in BHU was amazing. I would return to my room past midnight after doing some experiments, and still find laboratories lighted in various departments such as botany, zoology and agriculture. People talked about research and publications. In retrospect, the quality of research may not have been great, but the research environment was outstanding. It was probably one of the few universities where research was considered to be an important aspect of academic life.

At the end of my first year of studies (June 1952), I found that I was one of the top three performers in the examinations. My final M.Sc examinations in physical chemistry were held in May 1953. I submitted my thesis after the examinations. A few weeks later, I learnt that I had obtained a first class with second rank in M.Sc (chemistry). The first rank had gone to a classmate who later worked in the Indian Tobacco Co.

BHU had an active Karnataka Association which organized many events. I took part in Kannada plays and other activities. Kengal

Kannada Play-Banaras

Hanumanthiah (Chief Minister of Mysore) visited BHU at that time and spent an afternoon with a few of us. He was an impressive person.

Banaras not only gave me a tremendous start academically, but also guided me spiritually. Banaras truly represents India. Banaras or Kashi is the city of light. According to mythology, it is the city where Lord Shiva had to try hard to find residence. Today, one sees dirt and refuse everywhere in the city. It was so much cleaner when I was a student. I enjoyed going in a boat on the Ganges and visiting various temples, in particular Kashi Vishwanath. I consider Kashi Vishwanath to be one of the greatest temples of India. It is a people's temple where anyone can go and worship the deity as one desires. He is the only god who allows to be hugged and massaged. Banaras gave me this extraordinary experience. Kashi Vishwanath and the Sankat Mochan (where Tulsidas wrote Ramacharit Manas) have been special to me for long, and Banaras remains close to my heart. It is indeed an eternal city.

Another aspect of my life that was influenced by my stay in Banaras relates to my interest in music. I would go to concerts organized in the music college by its Principal, Pandit Omkarnath Thakur. I learnt much about music there and listened to a number of great musicians including the great Paluskar, who passed away at a young age some years later.

Linus Pauling:
Rao's boyhood hero

Bismillah Khan frequently played shehnai in our hostel.

I read the book, *The Nature of the Chemical Bond* by Linus Pauling, for the first time, in BHU. I had borrowed the book from a faculty member and kept it unreturned for a year. I went through it over and over again and admired it. I was amazed by the way Pauling approached chemistry. I understood how the structure of atoms and molecules was important in understanding chemistry. Pauling had not only studied structures of chemical compounds but also of solid materials. Pauling became my hero after I had read the book.

It was in Banaras that I took the decision to pursue scientific research as my life's mission. It was there that I first got the glimpse of what chemistry can achieve and the type of chemistry I should do. I was so excited about research that I could not help talking about the work of great scientists. Some of my friends enjoyed taunting me by addressing me as "professor". My best friends in the Banaras days were K. Nagesh Tantry (who was a reader in science education in Mysore, and died some years ago) and S.N. Balasubrahmanyam (SNB retired as a professor at IISc, a few years ago). I was close to SNB's family as well. His father was a great doctor and was a patron of South Indian classical music. U.R. Rao (former chairman of the Indian Space Commission) was my contemporary in BHU. He studied physics.

I published two papers based on my M.Sc thesis in 1954. My first publication was with a lecturer from Indore (A.P. Saxena) who came to Banaras to do some research. Prof. Joshi asked me to work with him on a small problem and he published a paper with me in the *Agra University Journal of Research*. I published another paper on my own in a British journal.

After completing the M.Sc degree in June 1953, I was wondering about the future, of what to do and where to go next. Indian Institute of Science (IISc) was the obvious place, and I did go there for a few days. Soon I realized that physical chemistry in the Indian Institute of Science was not strong. Furthermore, IISc was not conferring its own degrees or providing proper stipends. I was advised by Prof. K.R. Krishnaswami to go abroad or to the new institute being set up near Calcutta (now Kolkata), the Indian Institute of Technology (IIT) at Kharagpur. Much was expected of the IITs which were envisaged by Prime Minister Nehru as the Indian equivalents of MIT and Caltech. Sir J.C. Ghosh who was director of IISc had moved to IIT Kharagpur as its first director and was looking for young research scholars. I was interviewed in Bangalore and was selected. I decided to go to IIT Kharagpur. The chemistry department there had many faculty members from IISc Bangalore. Most of them were lecturers. The assistant professor in-charge of the department (Dr. S.K. Bhattacharyya) was from IISc and worked on catalysis. Dr. M.S. Muthana (polymer chemistry) and Dr. M.V.C. Sastri (adsorption of gases) were also from IISc. I was to work on gas adsorption in relation to catalysis, my immediate supervisor being Dr. Sastri. I spent a few days in the adsorption laboratory in IISc before leaving for Kharagpur.

My stay in Kharagpur was interesting. More exciting than my research, was the introduction to Bengal. I loved Bengal for its literature, theatre and music and its people. IIT Kharagpur functioned in the old Hijli jail which had housed patriotic prisoners at one time. My initial efforts involved fabricating a gas adsorption apparatus. I had to spend endless hours cajoling the senior glass blower (Mr. Talukdar) to get it fabricated. I taught Kannada to his children who had come with Talukdar from IISc Bangalore and had taken Kannada as the second language in the local school. Dr. A.K. Bose (a lecturer in organic chemistry) who had come from MIT in the US once talked to me about the importance of infrared spectra of organic compounds. I had wondered what it was about. I started looking at research papers in chemistry journals, journals of physical chemistry and chemical physics in particular. I could not understand much in those papers. Many of the papers were on molecules, spectra, structure, dynamics and such topics. I did not have the background to understand them. I really wanted to work on molecular structure and spectroscopy. It occurred to me that if I stayed in IIT Kharagpur much longer, I would be wasting my time and learn little. I had to go to a place where chemistry of the modern variety was being practiced. It meant going to the United States. Many of my friends, in particular my close associate V.S. Ramachandran (who is a well-known cement chemist settled in Canada) and D. Venkateswarlu (a lecturer in chemical engineering, later professor at IIT Madras) encouraged me to go to the US. I decided to apply to the US universities (thanks to Sir J.C. Ghosh and Dr. Sastri who approved of the idea).

In late 1953, I learnt that Pauling was coming to the Indian Science Congress in Hyderabad during the first week of January 1954. I went to Hyderabad by train at my own expense (by unreserved third class) to see Pauling, on the pretext of presenting a paper. Unfortunately, Pauling did not show up since he was denied a passport by the US government. He had been declared a Communist by the American government. I was disappointed, but I saw Pandit Nehru, and the opening of the Regional Research Laboratory by him.

I received offers of admission and financial support from a few American universities including Columbia, Penn State and Purdue. I decided to go to Purdue. The decision was easy because there was a professor in Purdue (Professor Robert Livingston) who was investigating molecular structure by electron diffraction. I was interested in this topic since I had seen papers by Linus Pauling on molecular structure using the electron diffraction of gases. I got a passport from Midnapore with some difficulty, with D. Venkateswarlu providing the financial guarantee. Going to Midnapore from Kharagpur was no easy task in the hot and humid weather. One had to travel by bus. When it rained, an umbrella was required to avoid getting drenched inside the bus. I obtained a visa from the US Consulate in Calcutta after a medical check-up. I remember giving prints of all my fingers and swearing that I would not topple the US government.

When I told my parents about my plans to go to the US for Ph.D studies, they were happy, but were somewhat unnerved. My mother's only concern was that her only child was going far away, but she seemed determined to pray and wait. My father's immediate worry was how to find the money for my travel and other expenses. In his usual style, he calculated the exact amount required (Rs. 5500/- for travel and Rs. 700 for first month's expenses in the US) and promised to find it. (The exchange rate then was around 4 rupees a dollar.)

Air travel was not common then and I booked berths on boats travelling from Bombay to London (SS Corfu) and from London to New York (SS Queen Mary). My father got me some new clothes including a woolen top coat. My maternal uncle (Swamy) gave me a leather briefcase. My mother gave me a picture of Lord Vishnu, blessed by my grandfather. I touched her feet and also of my grandparents before leaving. I left Bangalore in late 1954 for Bombay (now Mumbai) with my father, my uncle Murthy and a few friends by train to take the boat to London. It was difficult to see my father crying when the boat left the shores of Bombay. It took 17 days to reach London where I spent two days in a hotel before taking the boat from Southampton to New York. It was a five-day journey. All through the boat trips, I was worried about the orientation exam that I was to take soon after joining the chemistry department at Purdue. I tried to study, little did I realize that the exam in the US would be of a different nature. When I landed in New York, I went directly to the YMCA on 35th street (as advised by an acquaintance who was in the US); it was the cheapest place to stay

($3.25 per day). I could reach Purdue University in Lafayette, Indiana by train or bus, and I opted for the Greyhound bus which cost less. I remember a funny incident at the Greyhound bus station in New York. At the coffee counter, I saw some round brown items that looked like vadas. I bought one and was disappointed by its sweet taste. I had not seen doughnuts before! I arrived in Lafayette the next day, early in the morning.

The orientation exams in the chemistry department at Purdue were full of objective questions and problems. I was not familiar with such questions, but did fare reasonably well. I was advised to take courses in various branches of chemistry. Since I wished to have physics as part of my studies, I could take physics courses as well. I had chosen a chemical physics programme with physical chemistry as the major subject and physics as the minor subject. Expertise in other branches of chemistry was equally required to pass the qualifying exams in the chemistry department. Many of the students advised me not to take the chemical physics programme since it would take a long time to complete the Ph.D degree. Before the first semester was over, a note that I had submitted to *Science* (based on some of the work that I had done independently in BHU) got published.

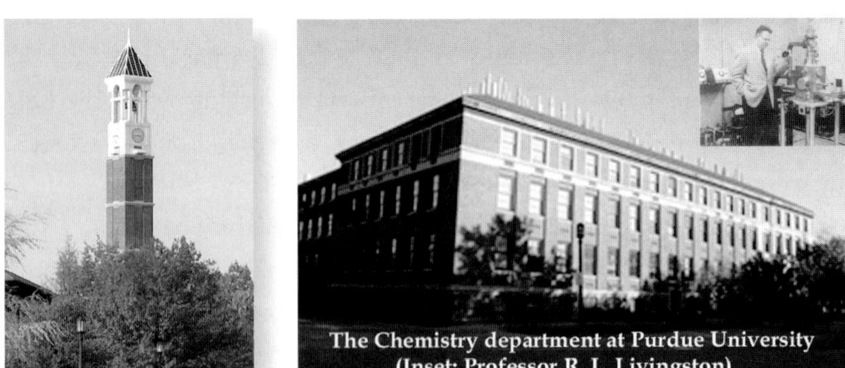

Purdue clock tower and chemistry department

I did not realize when leaving India that I would not be starting my Ph.D research immediately after landing in Purdue. To my surprise, I was a research assistant to a professor of organic chemistry (Prof. Eugene Lieber). He wanted me to study physical and spectroscopic properties of some of the heterocyclic compounds made in his laboratory. This gave me an opportunity to learn spectroscopy. I drafted my first paper with him just a few months after joining him. He took me to the National Meeting of the American Chemical Society at Minneapolis in September 1955. During the train trip from Chicago to Minneapolis, I was talking to a nice gentleman with a slight German accent sitting next to me. I found out later that he was the great Peter Debye (Nobel Laureate). I published several papers with Prof. Lieber, some even in organic synthesis. Most of the papers dealt with infrared and electronic spectra of molecules. Prof. Lieber was a kind person and was very good to me. When he left Purdue a year later for Chicago (130 miles away), I continued my association with him. I would stay with him in Chicago occasionally. He always had prunes for breakfast and would tell me how they were good for health and digestion. Even now, when I see prunes at breakfast time, I eat a few of them (and remember Prof. Lieber).

Prof. Herbert C. Brown — Rao's teacher

A year after joining Purdue, I completed the qualifying exams (in the four branches of chemistry) and started my Ph.D work with Prof. Robert Livingston (who belonged to the Pauling school). He was a fine person and wanted to build a new gas–phase electron diffraction apparatus. I got involved in the fabrication of the new instrument with two other students. Prof. H.C. Brown (Nobel laureate in chemistry, whose courses I had taken) was a source of great inspiration. He encouraged me to publish little notes and papers based on the research (mostly on chemical spectroscopy) that I used to carry out on my own, on ideas arising from seminars and courses that I attended. Prof. Yearian who taught me a course in the physics department allowed me free use of his X-ray laboratory and gave me the keys to it. This helped me to carry out X-ray diffraction studies of solids. The Purdue physics department had faced some misfortune with regard to the discovery of the transistor. Prof. Lark Horowitz and his student had observed transistor action (before the Bell Group), but did not know how to interpret it. If only they had gone to John Bardeen for advice!

Prof. Brown used to say, "If it is worth doing, it must be worth publishing". Faraday had said much the same many years earlier. In science, we "work, finish and publish". These statements have guided me all through my professional life. By the time, I started research for my Ph.D thesis on molecular structure by electron diffraction of gases, I had published several research papers on other topics. By then, Prof. Livingston had become Associate Dean and was hardly available for discussions. He told me that he expected an outstanding thesis from me and gave me the freedom to choose my own problems. This was a boon and I determined the structures of some interesting molecules. Prof. Livingston became a dear friend. He took me home for dinner and drinks regularly, and for lunch at the Sagamore restaurant in Union Building on Saturdays. He asked me to help his other students with their thesis work. I used the first generation computer for my calculations, one of the few to do so in the chemistry department. It was a massive computer full of old-fashioned valves. Transistors were still not being used in electronics those days, the transistor itself being a discovery of the early 1950s. What the big computer did for me then is now done by a tiny hand-held computer.

Linus Pauling visited Purdue to inaugurate the new chemistry building. I could show some of my electron diffraction results and explain how the structures of certain molecules that I had determined agreed excellently with his predictions. Pauling cited some of my results in the third edition of *The Nature of the Chemical Bond*.

Research at Purdue was exhilarating and I became preoccupied with publishing research papers. This had some effect on my course work. I would opt to get a B grade instead of an A grade in some courses because I was too busy with research to study for the exams. However, I did sufficiently well for the department to offer me a special research fellowship meant for the most outstanding student. I was elected as student representative in the departmental advisory committee. Considering that there were 350 Ph.D students in the department, this was an honour. I had a large circle of friends, particularly organic chemists, many of whom I helped with problems in physical chemistry and in the interpretation of spectra. When I appeared for the final Ph.D exam in January 1958, wearing a new suit (bought with $25 borrowed from a friend), the examiners just shook my hands and congratulated me. The exam was over in five minutes! I had just taken around 3 years to complete the Ph.D degree in chemical physics. We had a big party in the Chesterfield bar in town to celebrate the Ph.D degree. I remember Derek Davenport (a young assistant professor who hailed from London), toasting the Commonwealth with me.

During my stay at Purdue, my close friends were Sandy Kern (at Colorado University), Ed Kearns (who retired from DuPont Co. some time ago), Cal Kobayashi (from Hawai) and Harry Rubinstein (who has retired as a dean from Lowell University, Mass). A problem that I often faced during my Ph.D studies was the perpetual shortage of cash towards the end of each month. My scholarship of $150 per month was barely enough to make both ends meet. I would borrow 10 dollars from the dean regularly and return it on the first of the following month. There were only ten Indian students in Purdue when I first arrived, but there were close to 60 when I completed my Ph.D degree. The year I had joined Purdue, seven other Indians had also joined for Ph.D, but only three of us completed the degree. The Indian Association at Purdue was quite active and I took part in some of their programmes.

It was a satisfying experience to get a Ph.D degree, but more importantly to have published research papers on molecular structure and spectroscopy. To some extent, I had achieved the kind of research that I had dreamed of. As I was finishing my Ph.D degree, one of my classmates (who was working with Dr. J.M. Honig, an assistant professor in the department) asked me if I could determine the structure of the spectroscopically pure TiO_2 that he had prepared. He thought that it had the rutile structure. I found that it had the anatase structure. This led us to study the kinetics of transformation of anatase to rutile and gave me my first opportunity to publish a proper paper in solid state chemistry. After my Ph.D degree, I was appointed an instructor in the department with a much improved salary. I taught a graduate course in physical chemistry and a one-credit course on infrared spectroscopy. I started worrying about where to go next. Prof. Livingston worked hard on this problem. Prof. Lipscomb (who later received the Nobel Prize) wanted me to join him as a post-doctoral associate after 6–8 months since he was moving to Harvard from Minnesota. I could not wait that long. I had received an offer of a research assistant professorship from the University of Southern California. Fortunately, Prof. Kenneth Pitzer of the University of California at Berkeley offered me a research associateship around the same time. I decided to leave for Berkeley. Leaving Purdue was a moving experience. Prof. Livingston shed tears at the railway station.

University of California at Berkeley (UC Berkeley) is one of the most outstanding universities, and the Chemistry Department there ranks first amongst the universities of the world in research performance. This department was established by G.N. Lewis, who is considered to be the father of modern chemistry. G.N. Lewis was the first person to propose the idea of the chemical bond — When he placed two electrons between

Berkeley clock tower with G.N. Lewis inset

two atoms in 1916 and stated that it was a chemical bond, he changed chemistry. He trained a number of students, many of whom received Nobel Prizes. Glenn Seaborg, Melvin Calvin and Harold Urey are some of his well-known associates. Lewis made thermodynamics part of chemistry, discovered heavy water, created a new concept of acids and bases and contributed to the foundations of spectroscopy, photochemistry and electrochemistry. The idea of the chemical bond or any of the above contributions could have been the basis for awarding him a Nobel prize. Unfortunately, he did not receive a Nobel Prize. Other founders of new areas in science who have not received the Nobel Prize are John Slater in quantum mechanics and Henry Eyring in chemical dynamics. Lewis was found dead in the laboratory in 1946 and there was a rumour that he had committed suicide. The story is as follows. G.N. Lewis had asked one of his Ph.D students, Michael Kasha, to record the spectrum of a compound in a solvent of high dielectric constant. The solvent was liquid

Prof. K.S. Pitzer with whom Rao worked

hydrocyanic acid (HCN). Kasha was not enthusiastic about making liquid HCN which is highly poisonous. Lewis told Kasha that he himself would prepare liquid HCN and asked him to go for lunch. When Kasha came back from lunch, there was a broken ampoule of HCN and G.N. Lewis had died sitting in the chair. Did Lewis make it look like an accident? We will never know. It is noteworthy that Linus Pauling dedicated his classic book, *The Nature of the Chemical Bond*, to G.N. Lewis.

Kenneth Pitzer with whom I was to work had been hand-picked by G.N. Lewis as a member of the faculty at Berkeley, after Pitzer finished his Ph.D degree. Pitzer became a member of the U.S National Academy of Sciences at a very young age. He had made fundamental contributions to chemistry even as a young man, the barrier to rotation around the carbon–carbon single bond being one of them. He was an expert in spectroscopy, thermodynamics, electrochemistry, quantum chemistry and a variety of other areas, and worked on these areas in parallel. He believed in Lewis' definition of physical chemistry. "Physical chemistry is anything that is interesting". It was a pleasure to be at Berkeley to work with Prof. Pitzer, who was then the Dean at the College of Chemistry.

CNR Rao at Berkeley

UC Berkeley was an exciting place. One could feel the ghost of Lewis in the chemistry department. There were seminars by well-known scientists all the time. As a research associate, supported by the Radiation Laboratory (now Lawrence Berkeley Laboratory), I was paid a stipend of $625 per month. I lived in an apartment which I shared with J. Ramachandran (who was later a Professor at the University of California, San Francisco, and then the President of Astra Research Laboratories in Bangalore). I carried out research in a number of areas including NMR spectroscopy and heat capacity of solids and published a few papers with Prof. Pitzer. I had to partly work in the low-temperature laboratory of Prof. W. Giaque (Nobel Laureate). I was allowed to carry out some independent work which I published in *Nature* and other journals. My stay in Berkeley was made memorable by frequent visits to beautiful San Francisco across the bay. I often listened to Jazz concerts by many great artists such as Louis Armstrong and Dave Brubeck.

It was common to see great scientists in the corridors of UC Berkeley. Oppenheimer and Edward Teller visited the campus frequently. Those were the days when hot television debates were raging between Pauling and Teller on nuclear testing. Somehow, Teller always came out better on TV. I used to feel miserable about this. Oppenheimer and most of the Berkeley professors were against Pauling's views on nuclear testing (today Pauling's views form the basis of the world doctrine). What is remarkable is that a few years later when Oppenheimer was being branded as a security risk to the US, the two people who came to his defense were Linus Pauling and Hans Bethe. Both had severe disagreements with Oppenheimer on the hydrogen bomb and other matters, but they swore that under no circumstance was Oppenheimer a threat to US security. Great statesmanship indeed!

The Maharaja of Mysore (then the Governor of Mysore), Jayachamaraja Wodeyar, visited Berkeley for a day. I took him around. He was really impressed when he saw the accelerators in the Radiation Laboratory on the hill. He was happy to speak to me in Kannada and was even more impressed that I knew a little Sanskrit.

I should mention my encounter with the then Chairman of the Union Public Service Commission of India in 1958. He was rude to the few of us who met him and told us how he did not think much of American

Ph.Ds. Even doctorates from Oxford and D.Sc's from India were getting a starting salary of Rs. 300 or 400 per month, he said. Imagine, he said, D.Sc is the highest science degree in India. Immediately, I thought that I should get a D.Sc degree from India. I bound some of my independent papers (outside my Ph.D work), wrote an introduction, and submitted it for the D.Sc degree of the University of Mysore.

In early 1959, I received a letter from my father asking me, for the first time, about returning home. He wrote that since I was out for almost five years, I should consider coming home as my mother was becoming anxious. This letter started to bother me. I corresponded with Vice-Chancellor A.C. Joshi of Punjab University (who had visited the U.S) and he offered me a readership in physical chemistry in Punjab University with a basic starting salary of Rs. 700 per month (around Rs. 1000, including allowances). I wrote a letter to the Indian Institute of Science enquiring whether there were faculty positions in physical chemistry. A few months later, I received a letter from Dr. S. Bhagavantam (Director of IISc) who offered me a lecturership in physical chemistry at a basic salary of Rs. 500 (around Rs. 720, including allowances) per month. Prof. K.S.G. Doss, who was the Director of the Electrochemical Research Institute at Karaikudi, was interested too and offered to pay me a salary of Rs. 1000/- per month. Some of my friends advised me to explore ICI (Calcutta) which offered high salaries. All of this happened quickly, when I had not yet decided whether to return to India or not.

My research was progressing very well in Berkeley and I had a credible list of research publications (totaling around 30). I was getting to be quite popular in the department. I had excellent academic friends like George Pimentel who was a great chemist as well as a popular teacher of undergraduates. I was getting to know people like Glenn Seaborg and Joe Hildebrand. Glenn Seaborg, despite being a Nobel laureate and Chancellor of the University, taught undergraduate classes. Joe Hildebrand lived upto the age of 104 and published four papers when he was 100 years old. The Berkeley campus was academically overpowering. In the Tuesday evening colloquia, one would see a galaxy of chemists in the front row, with many Nobel Laureates and members of the National Academy of Science amongst them. It was a daunting task to give talks at those meetings. I had managed to survive those occasions. How could I leave such a place? Prof. Pitzer was asking me if I could stay on for some more time. There were queries if I would be interested in a faculty position in a US university. I did not know what to do. My teachers, including Professor Livingston, felt that I would be better off staying in the U.S since I was comfortably assimilated into the American academia. Most of my Indian friends felt that I would commit professional suicide by returning to India which was in a poor state in all aspects. It was an agonizing period. Should I be concerned about my research career or about my parents and my country? It was a difficult period during which I had many a sleepless night. Images of my mother and India often came to me.

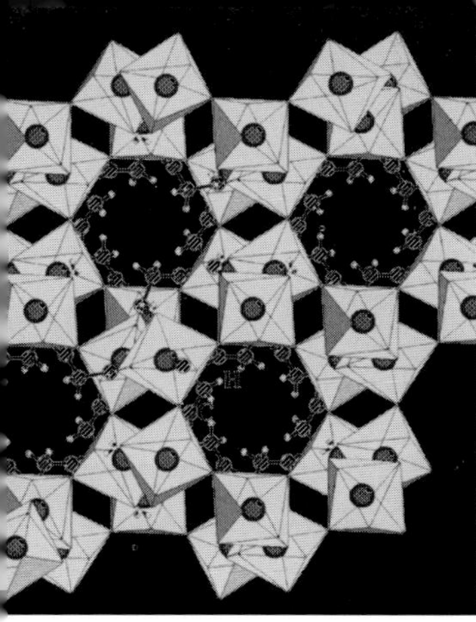

3

In Search of a Place and a Purpose

I decided to return home in 1959 since I thought that I could be of more use in India than in the US, and also because I wanted to make my parents happy. I reached Bombay by air and stayed there for two days to clear the luggage that was sent by sea through customs. After completing this horrendous task, I landed in Bangalore by a small Dakota aircraft from Bombay which arrived early in the morning. My parents, my grandparents and a few close relatives were at the small airport to receive me. It was an emotional welcome. It was good to be back in Bangalore. Bangalore had not changed.

I joined the Indian Institute of Science (IISc), which had offered me the lowest rank amongst the three offers that I had, and also the lowest salary. I felt that IISc would be the best place to carry out independent research. I visited the Institute and the Department of Inorganic and Physical Chemistry (IPC), the day after I arrived in Bangalore. The Institute was celebrating its golden jubilee. IPC is one of the two departments that was established when the Institute was founded. It was a small department (like most others) with five faculty members (including me). The department had recruited two Ph.D students to work with me. There were no mentionable facilities for the kind of research that I wanted to pursue. There were no spectrometers or X-ray diffraction equipment. There was a manual optical spectrometer. There was an old German X-ray generator with a tiny camera of 5.7 cm diameter. I had to plan my research with these meager facilities. I had brought some chemicals and other useful materials with me and started some simple work with one of the Ph.D students on certain organic molecules by examining their electronic transitions in different solvents. There was a single-beam infrared spectrometer in the physics department, but the head of the department (R.S. Krishnan) would not let me use it. He said, "why should chemists do spectroscopy?" He seemed unaware of what had happened in the science of chemistry. Since there was no infrared spectrometer, I managed to record routine infrared spectra of certain compounds here and there, in order to categorize certain group frequencies. One of those papers became a citation classic.

I was anxious to initiate research in solid state chemistry and had to think of a problem that could be tackled by the tiny X-ray camera. It could only give me a couple of lines in the diffraction pattern. There was better X-ray diffraction equipment in the metallurgy department, but the head of the department would not allow others to enter his domain. I found that the best option was to work on the transformations of TiO_2 (titanium dioxide) of different forms. Two lines in the diffraction pattern would suffice to study such transformations. I had a Ph.D student working on the transformations of TiO_2.

We wound furnaces with the heating elements and thermocouple wires obtained with great difficulty. We built other little pieces of equipment as well. I started an evening seminar in the department where students could present talks on various frontier topics in chemistry. I taught courses as part of the Ph.D programme. The head of the department (Prof. M.R.A. Rao) was a nice person and my colleagues in the department were not unfriendly. I could, however, see that they did not like my style. I used to mix freely with research students and would go for coffee with them. This was new to the department.

I met the Director, Dr. Bhagavantam, several times to appeal to him to procure one or two major instruments by using his influence to obtain foreign exchange. I would take catalogues and quotations to him. He was sympathetic but nothing happened even after two years. In spite of the acute shortage of facilities, I accepted more Ph.D students and ended up with six of them. I also started work on simple problems in solid state chemistry of rare earth oxides.

My best friends at that time were Satish Dhawan and Sivaraj Ramaseshan. Satish was a professor of aeronautical engineering and Sivaraj was an assistant professor in physics. Dr. A. Ramachandran, a Purdue alumnus, whom I had met at Purdue earlier, was in mechanical engineering. Prof. D.K. Banerjee who was the head of the department of organic chemistry was a great source of encouragement. He had obtained his D.Sc degree from Calcutta University, working on his own during the time of Acharya P.C. Ray. He enjoyed conversation and good food. The age difference between Prof. Banerjee and myself did not matter and we became good friends. He gave me a student from his department to work on problems in reaction mechanisms and spectroscopy. I extensively used the tiny infracord that he had managed to obtain. Prof. Banerjee was an outstanding organic chemist who had carried out difficult stereospecific and regiospecific synthesis of steroids and other complex organic molecules. I travelled a few times to the National Chemical Laboratory (NCL) in Poona (now Pune) to do library work and record infrared spectra of a few compounds. I overcame the little pin-pricks and intolerance by absorbing myself in writing a book on ultraviolet and visible spectroscopy of molecules (which I had planned on even before I left Berkeley). There was no book on the subject that used modern quantum mechanical notations although the technique was widely being used by chemists. During the course of writing, I needed a reference book which was in the Raman Research Institute. When I made a request for it, Prof. Raman said that he would find it for me. A day later, he personally came to my room to give the book. I completed my book on ultraviolet and visible spectroscopy by the middle of 1960. The department would not provide the postal charges to mail the manuscript to the publishers in London.

During my visits to NCL, I had the opportunity to meet chemists such as S.C. Bhattacharya and A.B. Biswas. There were also a few solid state chemists in NCL. I wished that I had some of the instruments that they had. I got to know Dr. K. Venkataraman who was the director of NCL. He was a veteran organic chemist working on dyes. The first time that I met him, he mentioned that he could offer me a junior scientific officer's post (starting salary of Rs. 350 per month), and I could do spectra of the various dyes that he had. Two years later, when I participated in a symposium that he had arranged in Shillong, he offered me a better position. He was a polished individual, but I did not feel that my future was in NCL or in a laboratory of the Council of Scientific and Industrial Research (CSIR). At the Shillong meeting, I gave a talk on electronic spectroscopy of molecules. I got to know Prof. T.R. Govindachari who had established a great school of organic chemistry in Presidency College, Madras. The most enjoyable aspect of the Shillong meeting was the long train ride from Bangalore to Kolkata with Sivaraj Ramaseshan (who spoke on crystallography at the meeting).

I was quite outspoken in the Institute and some of the senior members of the faculty may not have liked the way I asked questions in seminars. An incident that I remember was when Dr. M.S. Thacker, who was the Director-General of CSIR, gave a lecture on science in CSIR at the Institute. I was sitting with Dhawan and Ramaseshan in the front row. At the end of his lecture, I had to remark that most of what he mentioned did not represent modern science. At least in chemistry, which was one of the main areas of CSIR, I did not see trends which were futuristic. Many people were a bit shocked by my intervention, but Satish and Shivaraj enjoyed it. I got to know Dr. Thacker later and there was no problem between us. In 1960, a low-energy conference was organized by the Department of Atomic Energy at IISc. During that conference, I met Dr. Raja Ramanna for the first time. I gave a talk on

TiO_2 transformations in a session chaired by Prof. G.N. Ramachandran (GNR). After my talk, he commented "I thought that you were a chemist!" I had admired GNR from the days of my youth, for his originality and brilliance. GNR was basically a simple person and child-like in many of his reactions.

Indian Institute of Science had its share of unique characters and personalities, some of whom I knew personally. There was M. Srinivasiah who had established the first fermentation laboratory. He was just a B.A, but outstanding in his work. He was known for his acidic tongue and had a nickname for everyone. Once he asked me whether I had met the blind fellow. I had to check from various sources to find out who he was referring to. There were gentlemen like K.R. Krishnaswami (KRK), head of the IPC department. He was a D.Sc (London) and had a style about him. I remember stopping by the department when I was visiting Bangalore from IIT Kharagpur. There was KRK getting into a taxi with my old friend V. Srinivasan. KRK was taking Srinivasan out lunch to celebrate his appointment as a junior research assistant. Once, KRK was approached by a clerk for a loan of 10 rupees. When the clerk said that he would return the money as early as possible, KRK gave him 10 rupees and remarked "You have the audacity to promise that you will return the money" before sending him away. In a student seminar in the IPC department that I once attended in the early 1950s, the student started his seminar with the words "The chalcopyrites of Chitradurga" (the mineral on which he was doing beneficiation.) He stopped, and then again said, "The chalcopyrites of Chitradurga". This went on for the third time and the student choked. KRK who was sitting in the front row stood up and said "Ladies and gentlemen, I think that we have listened to whatever the speaker had to say on the subject," and closed the seminar. There are many other amusing stories of the old IISc, but I refrain from going into the details for the sake of brevity.

My return to India from the US was for another important reason — a personal one. I wanted to find a nice, educated companion for life. I

Indu and Rao on their wedding day

did not have to wait for long. I knew about a brilliant, attractive girl, Indumati, who was the daughter of Mr. and Mrs. B.S. Narayanaswamy. She was a B.A (Honours) in English literature. Narayanaswamy was from the Bapu family and was a first cousin of my grandmother. He was educated in Manchester and was a senior engineer in the Mysore public works department. Narayanaswamy's father (Bapu Subba Rao) was one of the first Indian professors of Mysore University.* Indumati's mother (Leela Bai) was the daughter of a health officer in Madras and granddaughter of Karpur Srinivasa Rao, the great engineer who built the Krishnaraja Sagar dam (Brindavan gardens) in Mysore. My grandmother was very enthusiastic about such a relationship. I met Indumati in December 1959. (I had last seen her in the summer of 1952 when I was visiting Bangalore from Banaras. We were both in a movie theatre watching Gone with the Wind. She was with some of her cousins.) Our families met too. I wanted to ask Indumati whether she would consider marrying me. Since there were no easily accessible phones at that time, I had to call her from the departmental office after requesting the office staff to step outside for a couple of minutes. Whatever I told her must have been audible outside and everyone seemed to know the result of the phone call.

Indumati (Indu) had innumerable relatives in Bangalore. One of her uncles (Bapu Raghunatha Rao) was the officer-in-charge of elephants of the old Mysore state and another (B.S. Raghavendra Rao) was a civil

* He taught and advised great literary personalities such as D.V. Gundappa, Masti and B.M. Sri.

servant (finance secretary) in the Mysore government. The most interesting person was Dr. B.A. Rao, an expert in malaria, who had graduated from Harvard University in 1930. Another uncle of hers, B. Venkoba Rao, was the principal of the Agriculture College. Dr. B.A. Rao was a great man and was personally responsible for eradicating malaria in Mysore state and elsewhere. He is known to have brought a species of fish from Europe which would eat up mosquito larvae, and to have introduced the species in ponds and lakes all over. He served as the first director of the Institute of Malariology in Delhi and later in WHO. Mr. Raghunatha Rao, the elephant doctor, had lots of elephant stories to tell. He had a unique personality and took pride in the fact that he knew all the elephants in Mysore state. They all saluted him and obeyed him instantly. He later served the Maharaja of Jaipur. A characteristic of the Bapu men was their high integrity. They were strict about spending public funds and would not accept hospitality from any one when on official trips. If they had to eat with someone they paid for it. My father was similar. When he visited schools in interior Mysore state for inspection, he had to eat in a teacher's house out of necessity (there were no hotels and restaurants in most small towns). He always gave some money to the teacher.

Indu had many aunts, some of whom stayed in Bangalore. She had a large number of cousins and was close to some of them. Bapus were a close-knit family and the extended Bapu family had members who worked in all possible sectors. Our marriage was celebrated on May 15, 1960 in Bangalore in the presence of all the relatives. We lived in a house on 13th Cross, Malleswaram, not far from IISc. A few months after our marriage, my book entitled *Ultraviolet and Visible Spectroscopy* was published by Butterworths, London. The book was translated into several languages soon after. Our daughter, Suchitra, was born on March 30, 1961 in Bangalore.

Social life in the IISc campus was limited. The only people whom we visited were the Banerjees, Dhawans and Ramaseshans. Indu visited her relatives in south Bangalore regularly. Occasionally, we went to Cubbon Park or somewhere else for coffee or dosa. I used to visit Ramaseshans in the IISc Campus during the lunch hour when we commiserated together. It was during this period that I met G.N. Ramachandran who was not in the best of health. He was being treated by a distant relative of my wife. GNR had solved the structure of triple helical collagen by then.

On a morning in 1961, I received a brown envelope from the University of Mysore. The University had awarded the D.Sc degree for my thesis. It had taken three years for the University to get the thesis examined. It was common then to wait for 2 to 3 years to get a research degree from a university. Fortunately, I had a Ph.D degree. I started work on a major book on chemical applications of infrared spectroscopy in 1961. Academic Press, New York, had agreed to publish it.

In January 1961, I received an invitation from Dr. Vikram Sarabhai to give a talk in the physics section of the Indian Science Congress session in Roorkee. I was to talk on the use of spectroscopy in chemistry. Pandit Nehru inaugurated the session. I saw the great S.N. Bose in the front row arguing about something. It was a nice occasion for me to meet a number of spectroscopists and other scientists. It was then that I learnt about a new Indian Institute of Technology (IIT) coming up at Kanpur with the support of American universities. I also found that there was a first-generation NMR spectrometer in the physics department of Aligarh Muslim University. I used the instrument to carry out some work with the Aligarh physicists (P. Venkateswarlu and B.D.N. Rao).

An important event in Bangalore during this period was the celebration of the centenary of the great visionary, Sir M. Visveswaraya. Every few weeks before the centenary, there would be a rumour that Sir M.V had passed away. On one such occasion, C.V. Raman went with his wife to Sir M.V's house to express condolence. When he knocked on the door, Sir M.V opened it. A shocked Raman recovered quickly and just blurted "I thought that I should just say hello to you" and left immediately.

The Indian National Congress had its session close to IISc in the palace orchards (Sadashivanagar) area in June 1960, it was attended by Pandit Nehru and other leaders. There was a procession with the congress president (Neelam Sanjeeva Reddy). I went to see the procession (I believe along with Ramaseshan and Dhawan). We could see Pandit Nehru a few yards away. The chariot drawn by bullocks came down the road. When it came close enough to where Nehru was standing, Sanjeeva Reddy stopped the chariot and ran to Mr. Nehru to touch his feet. We could hear Nehru saying "get back to your chariot".

Three years after I had joined IISc in 1962, three of my students submitted their thesis for the Ph.D degree of the Institute. Around that time, I got an invitation to present a talk at an international conference on spectroscopy in Washington D.C. I decided to go. Since I was going to the US anyway, I wrote to Purdue university which then offered me a summer faculty position to teach a course in spectroscopy. Just before leaving Bangalore, my grandmother passed away. The binding force of my mother's family was gone. She would tell everyone who visited her before she died that her grandson was going again to America. I left Bangalore in May 1962 with eight dollars that I was allowed to take as foreign exchange. Fortunately, some friends received me in New York. The main purpose of my trip to the US was to collect speciality chemicals and minor items that I would require in the laboratory in Bangalore. After Purdue, I spent four months in Chicago to carry out some spectroscopic research, and returned to Bangalore in November 1962. Just as I was leaving the US, the Cuban crisis occurred. Indu and everyone else in the family were worried whether I would be able to return easily. As luck would have it, my Air India flight had to force land (due to a fire alarm) somewhere in the Middle East and I arrived a day late in Bangalore, much to the relief of all concerned!

It was good to be back with Indu and Suchitra. We had a small Morris minor car and life seemed quite reasonable. I enjoyed moving around Bangalore in the little car with Indu and Suchitra. Suchitra used to say "Let us go for a walk in the car". I gave a short car ride to her as a routine every morning before going to the Institute. The atmosphere in IISc and elsewhere was, however, depressing. No one seemed to worry about the facilities. By common consent, there was no future for anyone. There were many in IISc who retired as lecturers after having served the institute for 30 years or more. There was even a person who retired as a research assistant after having guided many research students. I was seriously worried as to how I could continue in IISc much longer without being able to carry out the kind of research that I really wanted to. It was near to impossible to accomplish what I wanted to do at IISc. I decided to leave IISc and go somewhere else. Most of the people that I knew in places like the atomic energy laboratories and Tata Institute of Fundamental Research were patronizing and unhelpful.

When I was in the US during May–November 1962, many of the professors there had asked me whether I would be interested in an associate professorship in New York or elsewhere. I had requested that they wait for a few months before I could decide. I wanted to discuss this matter in detail with Indu when I returned to Bangalore. Poona University was looking for a professor at that time to head its chemistry department, in place of Prof. S.K.K. Jatkar who was to retire. I had been nominated as a candidate and was called for an interview in Pune. After the interview, one of the members of the selection committee (Dr. B.D. Tilak, Director of NCL) told me that I could prepare to come soon to Pune. He also wrote to me to ensure that I would take up the job in Poona University, but I did not receive any communication from the university even after several weeks. I learnt later that H.J. Arnikar was appointed for the post. He was a lecturer in Banaras when I was a student. Providence had helped me by denying the professorship in Poona University. Fortunately, the new Indian Institute of Technology

(IIT) at Kanpur was looking for young faculty members. The institute was supported by a consortium of US universities which included MIT, Caltech, Berkeley, Princeton, Purdue, Ohio state and so on. There was promise of substantial aid for procuring scientific equipment. I was asked to appear for an interview in IIT Kanpur in early 1963.

I went to Kanpur for the interview by train and was shocked by the state of the city. The Kanpur railway station can dishearten even the most determined enthusiast. The campus of IIT Kanpur (IIT/K) was still in the making. After I talked to the Director (Dr. P.K. Kelkar), the Deputy Director (Dr. M.S. Muthana, earlier at IIT Kharagpur), and other members of the committee, I returned to Bangalore. Within a few days, I received an offer of an associate professorship in chemistry. Two others (M.V. George and P.T. Narasimhan) had also been offered associate professorships in chemistry. I left IISc in April 1963 to go to Kanpur. Satish Dhawan who had just become director of the Institute was unhappy to see me go and said that he would get me back sometime soon. Ramaseshan and T.R. Anantharaman (a fine physical metallurgist) had left IISc for other institutions a few months before me. Sukh Dev (an excellent organic chemist) had left the Institute much earlier to work in NCL, Pune.

Although I left IISc, I must mention in retrospect that I am so glad that my first academic position was secured there. It is extremely important to pick the right place to start one's career. It takes only one early mistake to spoil one's entire scientific career. This is partly because of the limited number of good institutions in the country. Some of my friends who were truly outstanding got lost by joining the wrong place to start their careers. The greatest virtue of IISc is the extraordinary academic freedom it provides. A lowly lecturer can have a larger research group than a mighty professor. (I am fortunate that I have always been a part of it.)

I went alone to Kanpur by train to join the new IIT. I was provided a small apartment in the campus that was far away from the city.

The Institute itself continued to function in the Harcourt Butler Technological Institute (HBTI) in the city. The HBTI building had primitive facilities and there was no place even to enjoy a cup of tea. For the first few months, we depended on wayside hawkers for samosas and tea at lunch time. I shared an office with many others and started looking at what I should do in the next few months. Derek Davenport, a chemical educator from Purdue, was then a visiting professor. Both of us went around the city in a bicycle rickshaw looking for glassware and other items for the freshman general chemistry laboratory that was to be initiated in July (in the workshop building in the new campus). We ordered some laboratory furniture from the local carpentry shops. I was to teach the freshman chemistry course to start in July. After a couple of months of stay in Kanpur, I brought my family, car and belongings from Bangalore. We lived in the apartment at the IIT/K campus through the hot summer. When the new academic session of 1963 started, I began giving lectures to the new undergraduates in a make-shift lecture hall. I would be dripping wet with perspiration after the lecture. The laboratory classes were held in the workshop, but the experiments that we had

IIT/K Kelkar Library with Kelkar inset

planned were quite interesting. How we survived those hot days still surprises me. There were no facilities worthy of mention in the campus, and one had to travel to the city several miles away to buy salt or sugar. Going to a restaurant was not an option because there was none. Kanpur offered little to the young faculty families who started arriving at the campus. The students were all in a hostel and there was little entertainment. It was a campus without roads or lights. I remember the evening when I found my car sunk half-way in mud. It required enormous effort to pull it out.

Those days at the IIT/K campus were also exciting. Although there were no amenities, there was hope and idealism. We started working on a new undergraduate curriculum with a core programme during the first three years. Lectures in the first two years were associated with tutorials. The tutors attended the lectures, immaterial of their rank. I was an enthusiast when it came to undergraduate teaching. Chemistry was always voted the best first-year course by the students. I was part of a committee to plan the campus buildings as well. More importantly, I was a member of the core curriculum committee. One of my close associates in the committees was Jagdish Mahanty of the physics department. He was a good friend and our families were close. He is probably one of the few friends that I have had with whom I rarely agreed, but this did not affect our friendship. Jagdish's wife Saroj was a dear friend of my wife and they met every day. Saroj was a wonderful human being and often narrated horror stories of the partition days that she experienced while leaving Lahore for India. I shared Dindigal cigars that Jagdish would procure. We also shared good tobacco for our pipes. (I stopped smoking in the late 1970s although I enjoyed smoking a pipe.) Unfortunately, both Jagadish and his wife Saroj who had migrated to Australia passed away a few years ago.

I started research with my students within a year after reaching Kanpur. We had a few modern spectrometers and other instruments that were set up as central facilities. We began to order chemicals and instruments through the American programme and slowly accumulated most of the items required for research. The American consortium helped IIT Kanpur in many ways. The first head of the American programme was Prof. Norman Dahl of MIT. He and Dr. Kelkar got along famously and this helped the Institute in a big way. The faculty members from the US universities were catalytic in giving the right shape to curricula. They also helped in the recruitment of faculty members, besides in the procurement of equipment from the US. The quality of the American faculty was somewhat variable. Some were outstanding as typified by Lou Smullin (MIT), Charlie Dryden (Ohio State) and Joseph Meyer (Berkeley). There was a massive faculty recruitment programme, and brilliant young faculty started joining the Institute from all over the world. It was a young campus. We were the first to get a computer in an academic campus. It was a festive day when the IBM computer arrived with all of us helping to shift it to a temporary laboratory. The computer arrived by plane all the way from the US, but from Kanpur airport to the campus it was transported by a

Rao at the Indian Institute of Technology/Kanpur

bullock cart. The crates were upside down! IIT/K was the first institute in India to impart computer education to students as well as to scientists and engineers in academia and industry.

By the end of the first year of stay in IIT/K, I was formally asked to be the head of the chemistry department. I was barely 30 years old. I had the co-operation of all the faculty, specially of my fine colleagues, M.V. George and P.T. Narasimhan. I was promoted to full professor soon after. I was one of the first five full professors of the Institute. In running the department and contributing to the Institute in various ways, I had to swallow minor insults and criticisms, and had to put up with the idiosyncrasies of individuals — I did not want minor differences to hamper the progress of the department or of the institute. IIT Kanpur emerged as the best equipped educational institution by the mid 1960s. In the chemistry department, we were keen on initiating Ph.D and Master's degree programmes, and started admitting post-graduate students from 1964. The course work was rather demanding both for the students and teachers. I taught the freshman chemistry course every year and it was a most enjoyable experience. I would perform all kinds of demonstrations for the young students. Besides the freshman chemistry course, I also taught physical chemistry for M.Sc students as well as specialized post-graduate courses on thermodynamics and spectroscopy.

By 1965, I had students working in areas related to spectroscopy and molecular structure, and a few in solid state chemistry as well. I could not, however, carry out experiments of the kind that I wanted with the available facilities. There were no dedicated instruments that I could use on a full-time basis. Most instruments were considered to be central facilities and this restricted the scope of research and the nature of problems that one could take up. Yet, it was a far better situation than the one I had faced at IISc. What made IIT/K different was the way we worked. Pursuit of excellence was a phrase coined there at that time. We had a great director in Kelkar who was good to young faculty and gave the freedom to plan and execute good ideas. He did not interfere with the functioning of the departments. Discussions with him were always about the philosophy and ideals of the Institute, and about ideas and plans that would make IIT Kanpur great. In tackling issues and dealing with people, he exhibited great sensitivity.

Dr. Kelkar used to tell us interesting stories and anecdotes some of which imparted much wisdom. He used to compare Saraswati (the goddess of learning) and Lakshmi (the goddess of wealth). It is easy to appease Lakshmi, he would say. If one prays to her she bestows some money. You pray some more, she gives more. Saraswathi is not so easy to please. Even when one prays to her constantly, she does not oblige. For her to grant something meaningful, one has to surrender oneself to her. It is only through complete surrender to Saraswati that one can become her blessed worshipper. One may also get rich eventually through her blessings. This message was similar to what my mother used to tell me. "Let all your gains be through knowledge. Money can be here now, but gone in a second. Knowledge and wisdom are permanent".

Towards the end of 1963, my book on infrared spectroscopy (700 pages, Academic Press, New York) was published. In July 1964, I received a letter from Prof. C.V. Raman stating that he liked what I had written on spectroscopy and that he would like me to be a fellow of the Indian Academy of Sciences (see the letter reproduced on the next page). This was my first academic recognition.

In September 1964, our son, Sanjay was born. I had written another book with a colleague of mine on experiments in general chemistry for undergraduate students (published in 1965).

For nearly a five-year period from 1963 to 1967, I did not go abroad (except for a spectroscopy conference in Australia in 1965) and neither did I travel much in India. I was publishing papers on spectroscopic studies of molecules, hydrogen bonding and charge-transfer complexes. I succeeded to some extent in publishing a few good papers in solid state chemistry. I could broaden the scope of my research in this area slightly

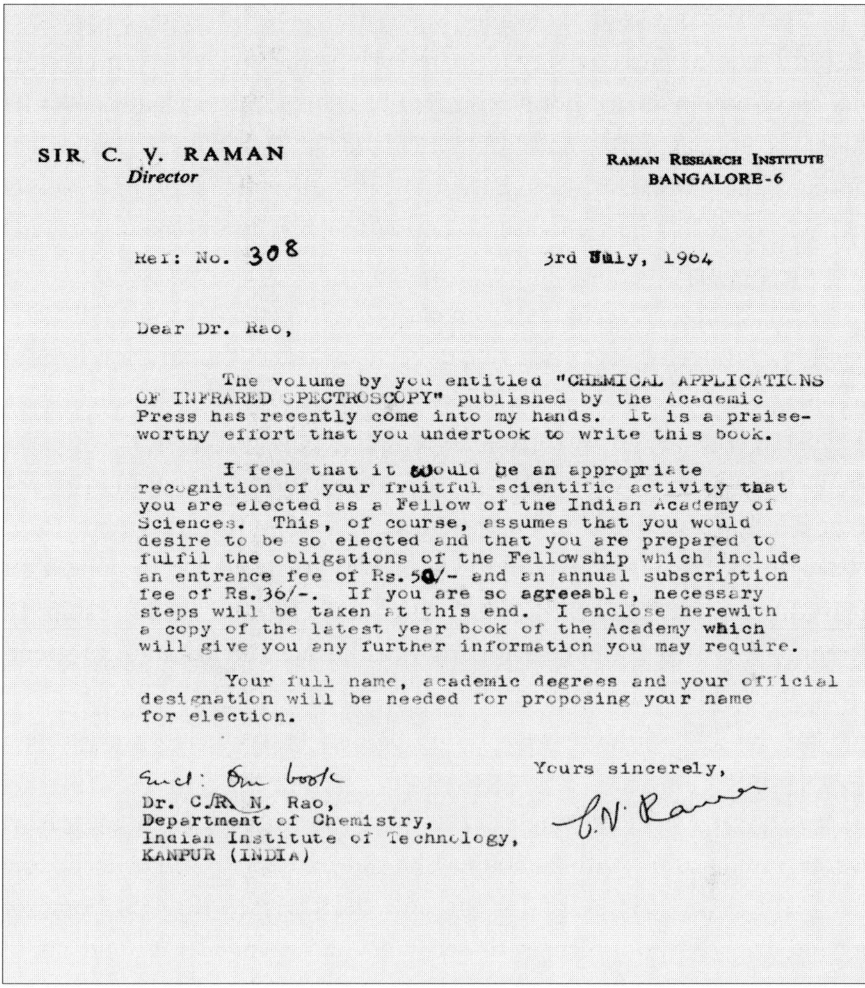

SIR C. V. RAMAN
Director

RAMAN RESEARCH INSTITUTE
BANGALORE-6

Ref: No. 308

3rd July, 1964

Dear Dr. Rao,

The volume by you entitled "CHEMICAL APPLICATIONS OF INFRARED SPECTROSCOPY" published by the Academic Press has recently come into my hands. It is a praiseworthy effort that you undertook to write this book.

I feel that it would be an appropriate recognition of your fruitful scientific activity that you are elected as a Fellow of the Indian Academy of Sciences. This, of course, assumes that you would desire to be so elected and that you are prepared to fulfil the obligations of the Fellowship which include an entrance fee of Rs.50/- and an annual subscription fee of Rs.36/-. If you are so agreeable, necessary steps will be taken at this end. I enclose herewith a copy of the latest year book of the Academy which will give you any further information you may require.

Your full name, academic degrees and your official designation will be needed for proposing your name for election.

Encl: One book

Yours sincerely,

Dr. C.N.R. Rao,
Department of Chemistry,
Indian Institute of Technology,
KANPUR (INDIA)

due to the improved facilities at IIT Kanpur. One of the important areas that I started to work on was transition metal oxides, because of the wide range of properties and phenomena exhibited by them. Hagenmuller in Bordeaux (France) and Goodenough in MIT (US) were two leading contributors to this area at that time. They were much older than me, but nearly became my contemporaries in the subject. I faced a constant dilemma as to what area I should pursue vigorously. Should it be spectroscopy and molecular structure, or solid state chemistry? Should I lean towards biology or materials? It was slowly becoming clear that spectroscopy might not be the area of choice since I would never be able to compete with those in the advanced countries on an equal footing. They would always have better equipment. Research grants available in India were modest at best. (The first grant that I received gave me two research scholars and Rs. 3000 for other expenses.)

I received an unexpected letter in early 1967 from the Faraday Society, London, informing me that I had been awarded the Marlow medal in recognition of my contributions to spectroscopy and solid state chemistry. The medal was given to a young physical chemist who was below the age of 35. This was the first international recognition that I received. The award was to be presented in Toronto in September 1967. I thought that it might be better to spend a year in the US to ponder on my future academic plans. Purdue offered me a visiting professorship. In September 1967, I travelled with Indu to Purdue (via Toronto) to spend a year there. Indu could get admission to the M.S. program in education. We left our two children with Indu's parents in Bangalore to facilitate Indu's studies.

Our stay at Purdue during 1967–68 was most useful. Indu got a M.S degree in education with honours. She did extremely well indeed. She met many of my friends at Purdue and elsewhere in the US. Some of my old friends were curious to see who had married me (and could manage me). I published several research papers with the Purdue faculty, especially in collaboration with George Honig and his students. I learnt much about the latest developments in solid state chemistry during that year. My stay in Purdue coincided with the excitement caused by certain metal oxides which showed transitions from the metallic to the insulating state on varying temperature or pressure. I worked on problems related to this phenomenon. More importantly, I came to the conclusion that I should concentrate on solid state chemistry and explore the synthesis and design of new functional materials. This realization was strengthened by the fact that the subject was in its infancy and there were very few practitioners in the world. Our stay at Purdue was made enjoyable by some of the Indian members of the university, specially A.K. Ramdas of the physics department (a student of C.V. Raman). We made many friends at Purdue, Henry and Paula Feuer being special. Both were professors and have remained close friends till today. I have two papers with Henry on the spectra of some compounds made in his laboratory. After participating in a Gordon research conference on inorganic materials, we returned to IIT Kanpur in September 1968 through Europe.

1967–68 was a momentous year in the US. Robert Kennedy was assassinated, soon followed by the assassination of Martin Luther King. Indu and I were on our way back to Purdue from San Francisco (where I had gone to attend the American Chemical Society meeting) when the assassination of King occurred. There were demonstrations of anger and frustration. Around the same time, President Johnson announced that he was not standing for election as President for the next term. Vietnam

CNR Rao receiving the Bhatnagar Prize in Chemical Sciences (1968)

war was in full swing and there were major protests on university campuses. The peace movement was specially strong in California.

We were welcomed back enthusiastically and affectionately by faculty and friends on the IIT/K Campus. We restarted our life on the campus and our children went back to the campus school. Life was enjoyable with frequent parties held by friends on the campus. I spent hours with my research students (to whom I wrote voluminous letters from Purdue). Boxes full of chemicals and other materials including heating elements, thermocouple wires, platinum sheets and wires, silver paint, etc., that I had brought with me gave a boost to my research. The Director wanted me to become the first Dean of Research of the Institute. I had no choice, but to take it. E.C. Subbarao was the Dean of Faculty. I was also involved in the interdisciplinary materials science programme with Subbarao. I organized winter schools in solid state chemistry in 1969 and 1971 with leading practitioners as lecturers. J.S. Anderson (Oxford), J.B. Goodenough, Leroy Eyring (Arizona) and J.M. Honig (Purdue) were some of them. Based on these lectures, the first book on solid state chemistry that I had edited was published in 1970 (Plenum Press, New York) and the second one in 1974 (Marcel Dekker, New York). The 1970 book entitled *Modern Aspects of Solid State Chemistry* is probably the first proper book on the subject.

The chemistry department at IIT/K had matured into an active and exciting place for education and research. It was probably the best chemistry department that the country has ever had. Many of the faculty members of the department whom we recruited have become well-known (Animesh Chakravorty, S. Ranganathan, (late) D. Devaprabhakara, D. Balasubramanian, J.C. Ahluwalia, P.K. Ghosh, G. Mehta, P.T. Manoharan, etc.). Animesh Chakravorty is one of the first modern inorganic chemists in the country while G. Mehta is a master of organic synthesis. The faculty was excellent in research and excited about teaching both undergraduate and post-graduate courses.

The first integrated 5-year M.Sc programme in chemistry was started in the department. This was actually equivalent to 4-year B.Sc + 1 year M.Sc, but the bachelor's degree was not awarded. The programme involved a few semesters of research as well. Alumini of the department have become leaders of academia in India, P. Balaram being most prominent.

I started taking serious interest in science education after 1969. Indu became involved in the Opportunity School devoted to poor children (children of maid servants and labourers) who were first-generation learners. She taught science and other subjects to the children in Hindi. The school was a great success and Indu fully identified herself with it. Opportunity School which started with a handful of students in 1969, has a few hundred students today.

There were international conferences on science education in India during 1968–71, organized with the support of the US National Science Foundation. I was an active participant in the conferences. During this period, I was invited to be a member of the chemical education committee of the International Union of Pure and Applied Chemistry (IUPAC). I organized a national meeting on chemical education and started a project on writing a General Chemistry text for the first college course in chemistry. A number of colleagues participated in this venture and K.V. Sane helped to edit the book. This book, published in 1973 (Macmillan), is still used widely. 1970–73 was a period when a large number of summer schools and winter schools were organized all over the country for school and college teachers. A National Council for Science Education (NCSE) was formed under the chairmanship of Prof. D.S. Kothari (Chairman of the University Grants Commission) to promote such programmes. R.D. Deshpande was the chief executive of NCSE. NCSE supported a science newsletter (Science Resource Letter) which I edited. D. Balasubramanian, M.M. Chaudhuri and others helped in bringing out the newsletter. We printed 15,000 copies of the newsletter and mailed it to schools and colleges. The difficult task of mailing was carried out by the children of the Opportunity School under the guidance of Indu. The school received a small payment for the service.

In November 1969, Prof. C.V. Raman turned 80. Vikram Sarabhai had organized the annual meeting of the Indian Academy of Sciences in Ahmedabad for the occasion. I had been asked to deliver a lecture. Prof. Raman was highly complimentary after my lecture on molecular charge transfer and said "you are a ball of fire". Raman gave a great lecture to school children on "Why is the sky blue". It was a memorable meeting where almost everyone in Indian science was present. During lunch one day, I heard Raman telling Sarabhai, "Vikram, you travel too much. Cut it down. It is not good for your health". G.N. Ramachandran was to give an evening lecture on molecular biology. He started the lecture nicely and choked after a few minutes. The lecture was cancelled. I saw this happen again some months later. I really felt bad to see GNR this way.

I met Professor Raman for the last time at the 1970 annual meeting of the Academy in Bangalore. He had asked me to present a lecture at the meeting. There were very few members, but Prof. Raman was there in the front row. I gave a lecture on the structure of water and the hydrogen bond. He was superlative in his comments after my lecture. In late November 1970, Prof. Raman passed away. He was a giant in science. Raman centenary was celebrated in 1989 in Kolkata where he had discovered the Raman effect. Rajiv Gandhi came specially for the occasion. I had the pleasure of organizing the celebration.

In 1971, I heard that I was being awarded the Bhatnagar Prize in Chemistry by the CSIR for the year 1968. (I did not even know that I had been nominated for the prize.) This gave a tremendous boost to my morale. I had to give the prize lecture in Delhi University. The lecture was chaired by Prof. T.R. Seshadri F.R.S, a very senior organic chemist of the country. I talked on the new metal oxides and the unusual phenomena exhibited by them. Many young people seemed to like the lecture, but Prof. Seshadri was rather curt and uncomplimentary in his closing remarks. Don't believe that all these kinds of materials and properties exist, he said. My wife who was in the audience could not believe what she had heard. I must mention that, by and large, members of the chemistry fraternity in India, including some of my colleagues, did not think much of solid state chemistry. I often used to hear of their ridicule, but I was committed to the subject.

Campus life in IIT Kanpur was truly fantastic. We had many parties in faculty homes and the relationship was close between the students and the faculty. The city being so far away and having so little to offer, helped the richness of campus life. We lived in a nice house on the campus (after the first year) which was a popular place for like-minded faculty to get together. I often think of the many fine people of IIT/K. Some of the American faculty added colour to campus life. There was great camaraderie and joy. There was harmless teasing and genuine laughter. I must admit, however, that this aspect of the campus life was somewhat confined to the western educated faculty members who formed the majority. I recollect many funny incidents as well as exciting moments of those days.

In a party hosted by an American professor, when a young staff member (who was a local recruit) was asked "Gin or Bourbon?", he answered, "No! I am a kshatriya". He had understood the query as "Jain or Brahmin?". We had many jokes about ourselves. There was a professor who was called "The Singh" because of his excessive usage of the definite article. He used to call the director "The Kelkar". One joke was about the Raos in the campus. "Someone called D.R. Rao and found that he had left to see Subbarao. They both had gone to meet Challapathi Rao. After finding Challapathi Rao to be absent, they had called G.N. Rao to find that he was with B.D.N. Rao."

My research group frequently met for group seminars and discussions. Equally common were the dinner parties that my wife and I gave to the research group at home. The first time we had such a party, all the food cooked by my wife and Kamla (our household help) was over in a giffy.

My wife had lots of vegetarian dishes, rice preparations, pooris and so on, but the students were ravenous. Hostel food did not satisfy them and they would go to the gym to work up an appetite before coming home for dinner. We really enjoyed having students and post-docs at home, a practice that we have continued over the years. Young Nihal Ahmed, my most loyal secretary, was always part of the student parties. Another enjoyable aspect of our life in Kanpur was our annual visit to the mountains. We actually got to experience the Himalayas there. We continue to travel to the Himalayas every year.

The IIT/K campus had its share of gossip and excitement. There were many rumours and family discords. What was specially tragic was the periodic suicides by undergraduate students. The suicides had generally to do with the students finding themselves inadequate and losing their self-esteem. A particularly terrible incident was when an undergraduate jumped from the roof of the five-storey faculty building on to the concrete ground below.

Dacoits were occasionally a sensation on the campus. Just a few weeks after we had started our life on the campus, dacoits entered our house. Fortunately, they stepped on Boregowda (a young boy who lived with us) who was sleeping on the floor. When he yelled, they ran away taking a few minor items. A very serious incident occurred several years later when dacoits entered the house of D. Balasubramanian (Balu) at night. When Balu tried to sit up, they hit him hard on his forehead. It was fortunate that the young child lying next to him or his wife was not beaten up. Balu was admitted to the local hospital. His head had swollen considerably and one could see one of his eyes bulging out of it. The eye surgeon suggested that the eye had to be removed if the other eye was to be saved. I had to take the decision on behalf of Balu and his family. Fortunately, Balu recovered well after a few weeks, but had permanently lost an eye (Balu is now director of research in the L.V.P. Eye Research Institute in Hyderabad). During his recovery at the hospital, Indu would visit him with milk payasam everyday.

One early morning, Chabbelal (husband of Kamla, our Nepali help at home) woke me up to inform that his friend, a Nepali security guard, at the campus branch of the State Bank of India had been murdered. I immediately drove to the bank to find that Muthana (Deputy Director) was already there. I saw the security guard lying in a pool of blood. I noticed the name "Babu Singh" written on the floor in blood (in Hindi). I pointed this out to Muthana and suggested that this may be a clue to the murder. He just laughed and said that I had read too many detective stories. Surely, a few days later, it was established that the chief security guard (Babu Singh) had entered the bank at night to rob the bank and killed the security guard (who had written the name in blood before dying).

Indu received an M.A in sociology from Kanpur University by studying privately, and at the same time increased her involvement with the Opportunity School. In 1971, the Government of India under Prime Minister Indira Gandhi decided to set up a National Committee on Science and Technology (NCST) to draft a science and technology plan for the country with Mr. C. Subramaniam (Minister for planning and science & technology) as the Chairman. There were nine members in the National Committee. I was surprised to be named a member of such a high-level committee. Raja Ramanna, V. Ramalingaswami, M.S. Swaminathan, M.M. Suri, B.D. Tilak and Hari Narayan were some of the other members. NCST was not too well-received by the heads of scientific departments and agencies, partly because the membership was restricted to non-official members who were not involved in running the government. Based on the recommendation of the NCST, Mrs. Indira Gandhi agreed to create the new Department of Science and Technology (DST) in the Government of India. After about a year, Dr. A. Ramachandran (then Director of IIT Madras) became the Secretary of DST. I worked closely with him on several programmes of DST (more about NCST and related aspects in Chapter 5).

Homi Bhabha visited IIT Kanpur for a day in 1966. There was a dinner given by Dr. Kelkar in the evening. Dr. Bhabha talked to me for a few minutes. Before he retired, he said that we should meet again after he returned from his trip abroad. That, of course, did not happen. The fatal crash of the aircraft with Homi Bhabha on the Alps shattered us all. Bhabha was a colossus and did much for science in the country. It is a pity that the country has not conferred Bharat Ratna on Bhabha posthumously. He was a gem. A few years later, Vikram Sarabhai visited IIT Kanpur, when he asked Dr. Muthana and myself to meet him in Bombay to discuss a collaboration between the Department of Atomic Energy and IIT Kanpur. Dr. Muthana and I both travelled to Bombay to meet him (in December 1971) and were staying at the atomic energy guest house on Malabar Hills when we heard of the tragic death of Vikram Sarabhai. He was on his way from Trivandrum to Bombay.

In 1971, Dr. Kelkar retired from the Directorship of IIT Kanpur to take up the directorship of IIT Bombay. This had a major effect on the institute. Kelkar was a fine human-being and an intellectual. He was respected by all members of the institute. Thanks to his support, we had built a great chemistry department. Towards the end of his stay, I would spend many hours talking with him on various issues. This was mostly after five in the evening, and my wife often complained to me about coming home late. Those discussions with him were probably part of my training for the future. There is no way I can adequately pay tribute to Dr. Kelkar. It may suffice to say that many of us owe our idealism, goals in life and growth in our professions to the Kelkar days of IIT Kanpur. Kelkar was missed by the IIT/K community. The Union Minister of Education asked me if I would take up the directorship of IIT Kanpur. It took me no time to refuse the generous offer. I was just too young (37 years old) to take up such a job and give up research and teaching. As I came out of the Minister's office, a bureaucrat who was present quipped "What is it that you want? What is your price?" I just told him that he would not understand. The Minister could not believe that I had refused the directorship of an IIT, and never failed to commend me on this decision whenever I encountered him later. Dr. M.S. Muthana, who was Deputy Director was appointed the Director of IIT Kanpur. He was a fine person and straight-forward to a fault. Dr. Muthana faced serious problems with the employee association and his relationship with Dr. Husain Zaheer (Chairman of the Board) was far from satisfactory. Some of the faculty members became somewhat political and there was an unhealthy polarization in the community.

In January 1974, I was awarded the title Padma Shri by the Government of India, but could not receive it personally, since Mr. C. Subramaniam wanted me to represent India in the science and technology discussions at the United Nations, New York. In May 1974, I received an invitation from the University of Oxford to be the Commonwealth Visiting Professor of Chemistry for the year 1974–75. The offer included a professor's salary and airfare for me and my family. Prof. J.S. Anderson was to be my host in the Inorganic Chemistry Laboratory (ICL) at Oxford. Both Dr. Zaheer and Dr. Muthana were delighted by the Oxford invitation and allowed me to be away for a year. Before travelling to Oxford, I attended a seminar in Delhi University. After my talk, Prof. Seshadri sent for me and said "I may not have been fair to you". I did not see Prof. Seshadri again. He passed away before I returned from Oxford.

Prof. Seshadri was a respected organic chemist. While I do not want to comment on his research work, I must express my admiration for his single-mindedness. Unfortunately, he was not generous to others who did not belong to his school, specially physical chemists like me. When he visited IIT Kanpur for the first time, I explained to him what we were trying to do in the chemistry department. His comment was "You want the moon! You won't get it". I could not help but remind him that people had already visited the moon. Somehow, our relations were never warm. Even with regard to my election to the fellowship of the Indian National Science Academy, he was not supportive. This does not prevent me from placing on record my appreciation of his determination to work in the laboratory till the last day. I feel bad that he was not well provided for in his old age. He deserved a lifetime professorship.

During the first 15–20 years of my career, I had my share of negative reactions and discouraging comments from some of the established professors. One or two scientists of the Bhabha Atomic Research Centre were particularly active in trying to damage me though subtly. Most of them talked nicely when I met them, but participated in a whispering campaign for a short period. In the hey days of superconductivity, one of my former colleagues tried to sow a seed of suspicion about my work. Fortunately, I had nothing to hide. While none of these affected my work seriously, I suffered silently. All this could be due to my style of talking or giving lectures. By the time I left for Oxford in 1974, I seemed to have gained some credibility. In fact, my Oxford visit changed many things for the better.

In early October 1974, I left with Indu and the children for Oxford. When I arrived in Oxford, I found that I had been elected a visiting professorial fellow of St. Catherine's college. Lord Bullock, the famous historian, was the master of the college. We lived in a nice cottage in Old Marston (a village just outside Oxford) and the children joined the local school. Oxford, with its colleges and spires, inspired us. We enjoyed walking through the streets of Oxford. Indu joined the education department of the university for higher studies. She worked on a thesis on the education of underprivileged children.

Prof. J.S. Anderson

ICL at Oxford was a great centre for solid state chemistry research. Prof. Anderson, F.R.S., was one of the early solid state chemists of the world and I collaborated with him as well as several others at Oxford. I forgot that I was a visiting professor and carried out a variety of experiments involving high resolution electron microscopy, neutron diffraction, photoemission spectroscopy and so on. I got to know Tony Cheetham, who had just become a lecturer and we wrote our first paper together then. Tony has remained a very dear friend of mine since that time. I got to know John Thomas who was then a professor in the University of Wales at Aberystwyth. We shared similar research interests and many of my students have since worked with him as postdoctoral fellows.

I remember the day after Christmas in 1974. I was working in my office when I heard some noise in the attic. I got a bit concerned since no one was expected to be in the laboratory during Christmas (the heating had been turned off and I was wearing a top coat in my office). I went to the attic and found someone wearing a top coat moving about. It was Prof. Anderson. He said that he was analyzing gallium in a compound made by one of the students. By the time we left Oxford, Prof. Anderson had become a personal friend. He enjoyed coming to our house for dinner. He specially liked almond halwa made by Indu. His reputation in Oxford as an uncommunicative and reserved person was certainly not right (Prof. Anderson worked till the last day of his life and died in 1988).

Indu received the Certificate of Education from Oxford. Unfortunately, she could not spend one more term there to get a B.Litt. Prof. Anderson was happy about the way I worked with so many people in Oxford. I published 17 research papers from Oxford, four with Prof. Anderson. I had started working on a book on *Phase Transitions in Solids*. McGraw-Hill, New York, had agreed to publish the book. We left Oxford for India in October 1975.

It was during my stay in Oxford that I met Prof. Nevill Mott (Nobel Laureate at Cambridge), the great solid state physicist, and discussed with him some problems related to metal–insulator transitions. We visited each other and wrote a paper together. On one of his visits to Oxford when he was talking to me in my office, I got a call from the vice-chancellor asking me whether Mott was there. The vice-chancellor shouted, "Ask him to rush. He has to receive an honorary doctorate in the next 30 minutes". Mott was a truly great scientist.

My Oxford experience was terrific and the college life was most enjoyable. (My association with St. Catherine's college has continued and I was flattered by my election as an Honorary Fellow in 2007. The present master, Roger Ainsworth is a dear friend.) We enjoyed going to the theatre in London as much as we loved the museums and art galleries. Oxford taught me to think differently about my area of research. It also revealed to me what would be required to carry out world-class research in solid state chemistry. I got the feeling that this could not be achieved in IIT Kanpur, considering the poor atmosphere that prevailed at that time. By late 1975, Dr. Muthana had given up the directorship and a person unworthy of such an academic office had become the director. The person would bargain with faculty members and even with vendors for minor gratification. I wanted to leave IIT Kanpur as early as possible. Where could I go if I left IIT/K which had given me so much? We had world-class students. Several people under me had received Ph.D and Master's degrees. I had research grants from the US National Institutes of Health (NIH), National Bureau of Standards and other agencies. I had good international collaboration, Ted Becker of NIH being one of my important collaborators. I had taught chemistry to innumerable undergraduates. We had a wonderful campus life.

I was again a worried man. Some of my friends in the US asked me to leave India and join a university there. They said that no one would complain that you did not try in India. I had a suggestion from Oxford about taking up a chair in chemistry. At that time, I was awarded a medal by the American Chemical Society which I was to receive in New York in September 1976. A dean of a major University attended my lecture to discuss salary and other aspects of a professorship that he was willing to offer.

With Prof. Satish Dhawan at IISc

When I returned from the US, I ran into Satish Dhawan. He first asked me if I would come to IISc and head the IPC department or the chemistry division. I told him that I was not interested in either position and was actually not looking for a job. Satish then asked me, "Why must you leave India? What will it take to get you back to IISc?". Such enquiries are rarely made by senior administrators. Satish was one of the finest human beings that I have known. I told Satish that I did not want to be part of any existing department in IISc. What I needed was an opportunity to build a new centre for research in solid state chemistry and materials science. He said that he did not have much money to give, but he would like to see me at IISc. Dr. Gurubaksh Singh had just taken over as vice-chancellor of the new central university at Hyderabad (of which I was a member of the planning committee). Gurubaksh wanted me in Hyderabad as the head of the chemistry school. When I said no, he became upset and did not talk to me for two years. Dr. A. Ramachandran (Secretary of DST) who was acting as the Director-General of CSIR wanted me at the National Chemical Labortory, Pune, as its Director in place of Dr. B.D. Tilak who was retiring. I did not want to become a director of a national laboratory. I remember meeting a senior professor of TIFR during those difficult days. He suggested

"Why don't you go as a professor to this new university in place X (which was not easily reachable by air or train)". I had to tell him that the university concerned was indeed the ideal place for him and not for a mortal like myself. In the mean time, I received the letter of appointment as professor with the maximum possible salary from the Indian Institute of Science.

4

A Long Journey

I decided to return to the Indian Institute of Science from IIT Kanpur and rejoined the Institute in November 1976. Satish Dhawan had promised me just a lakh of rupees and a secretary to start a new department devoted to solid state and structural chemistry. He had also told me that I could recruit two faculty members for the new department and requested me not to ask for a house in the campus since the housing situation was difficult. I was living in a fine house in the IIT Kanpur campus, but had to rent a house in Malleswaram paying almost half of my salary. I did not want to complain at that time or to get perturbed by extraneous factors. I just wanted to start building a new department devoted to my subject. The move to Bangalore caused considerable problems with the education of my children. Indu could not leave Kanpur with me since our daughter Suchitra was taking the higher secondary examination. I went to Bangalore by air from Delhi using my own limited funds. Sanjay came by train with a student from Kanpur to join the Central School. I brought our dog, Beatle, to Bangalore by air on one of my visits to Kanpur. He was small and was allowed to sit underneath my seat. The calmpose pills that I had given him, however, had no effect. Eventually, we all got together in early 1977 after Indu and Suchitra arrived in Bangalore. Suchitra joined a college in Bangalore.

I met some of my old colleagues in the chemistry division of IISc. The head of the chemistry division told me that there was absolutely no space to spare for my office or laboratory in the IPC department (to which I once belonged) or elsewhere. I could see run-down empty rooms with old bottles, but there was no point arguing with an unwilling colleague. It was a sad beginning. Fortunately, the head of the chemical engineering department (my old friend R. Kumar) offered me an office room and some laboratory space. The department of metallurgy gave me a small bay to set up a laboratory. I could convert an old solvent shed at the back of the campus to a laboratory. I could also convert a covered garage into a laboratory to house some instruments. The Department of Science and Technology of the Government of India had given me two major grants with about Rs. 80 lakhs for equipment and sufficient money for consumables, research fellows and other items of expenditure. I started some work with the equipment that I had brought from IIT Kanpur. Dr. Dhawan had told me that I could also build a Materials

Indian Institute of Science, Bangalore

Research Centre (MRC) with a new building making use of a special grant from the University Grants Commission, and procure some equipment for the Centre. This helped in a big way. I started looking for various spectrometers and other forms of equipment. I had to obtain import licences from the Government of India for getting the instruments from abroad (a terribly tedious process indeed!). Any instrument with a little computer in it required clearance from the department of electronics as well. With my grants, I could procure a sophisticated surface spectrometer which enabled me to establish the first such laboratory in the country. I obtained a variety of other instruments as well, including UV-visible and IR and EPR spectrometers. We installed the first laser Raman spectrometer in IISc (where Raman had spent many years). I, however, had to wait for several years more to have access to a good transmission electron microscope. Some of my Ph.D students and a postdoctoral research associate (P. Ganguly) from IIT Kanpur joined me in Bangalore. Dr. M.S. Hegde joined me as a post-doctoral fellow in surface science. I also recruited a few students from IISc.

I started exploring for funds for the Solid State and Structural Chemistry Unit. Within a short period, the University Grants Commission gave me a grant for a small building, running expenses and additional faculty. In a period of 2–3 years, the building of the materials research centre was completed. I had reasonable laboratory space, some sophisticated instruments and a fine group of people working with me. Good research seemed possible for the first time. My book on *Phase Transitions in Solids* with K.J. Rao was published (McGraw-Hill, New York, 1978). K.J. Rao (then at the National Aeronautical Laboratory) joined the Unit as a faculty member. J. Gopalakrishnan from IIT Madras and P. Ganguly were already members of the faculty of the Unit.

After some effort, Indu got a lectureship at the Teacher's College of the Mysore Education Society (MES) located in Malleswaram. She started a pre-primary school named Kishore Kendra which eventually became a high school. The school which she started with a handful of pupils now has over 1000 of them. Indu got interested in teaching geography and developed useful educational material on the subject.

G.N. Ramachandran had come to IISc from the University of Madras in 1971 to set up the Molecular Biophysics Unit, on invitation from Satish Dhawan. He was elected to the fellowship of the Royal Society in 1977. I was delighted that it had happened at last. He was 55 years old at that time. He should have been elected at least a decade earlier. GNR and myself had a common hero in Linus Pauling.

During 1976–78, I was a member of the Council of the Indian National Science Academy. Dr. B.P. Pal was President for two years and later Dr. Raja Ramanna became the President. Dr. Pal was a great plant breeder and agriculture scientist. He is also known as the rose man of India. The Council meetings during Dr. Pal's time were most enjoyable. He narrated joke after joke (many below the belt) and kept us in great humour. I should recount one or two to give the flavour. There was a conference on plasma physics in Ahmedabad which Prof. D.S. Kothari (Chairman of the University Grants Commission) was to inaugurate. In the introduction, Dr. Pal said, "some words mean different things to different people. For example, plasma has a different meaning to a doctor". Then he narrated this joke. "Two girls were in a doctor's office and one of them was very nervous. When asked why, she said that she was getting X-rayed. When she was assured that getting X-rayed was nothing much, she did not agree. She said that the previous month she had got ultra-violated". The last joke of Dr. Pal that I heard was when we celebrated his 80th birthday. In his reply to the felicitation, he mentioned the following. "There was a birthday function for a famous ophthalmologist. When he went to the hall where the ceremony was being held, he saw a huge painting showing the ophthalmologist coming out of an eye. He quipped, I am glad that I am not a gynaecologist".

During the 1977–80 period, I was a member of the University Grants Commission and also of the Scientific Advisory Committee to the Cabinet. The International Union of Pure and Applied Chemistry had asked me to be the chairman of the chemical education committee and had also appointed me as a member of the spectroscopy commission. These matters kept me involved in a few activities outside my academic commitments at IISc. I worked on various programmes of the Indian Academy of Sciences, becoming its first secretary, organizing national meetings of the Academy and other programmes. I worked with Sivaraj Ramaseshan closely on matters of the Academy and started new journals devoted to materials research and chemical sciences. I had to undergo surgery for upper abdominal hernia in 1979 in a local hospital, an experience that I do not like to remember.

In 1980, Suchitra completed her B.Sc degree and took up a one-year teachers' training course. In 1981, she got married to Dr. K.N. Ganesh who had just returned from Cambridge after his Ph.D degree.

Suchitra's wedding day (1981)

This was unexpected, but wonderful. Ganesh's father Mr. Nagappa was a mathematics teacher. Ganesh had come to see me a few times, like many other young people, to ask for professional advice about what he should do in India. He was toying with the idea of joining an industrial R & D laboratory. I advised him to go to the Centre for Cellular and Molecular Biology in Hyderabad, since he was interested in bio-organic chemistry. His visits to our house culminated in the marriage which we all rejoiced. My parents-in-law were in their element making arrangements for the wedding. It was a wonderful event in the family when all the four grandparents of Suchitra were healthy and active.

As time went on, I could improve the quality of my research and could train Ph.Ds in newer areas of solid state chemistry, surface science and molecular structure. I also broadened the scope of my research in solid state chemistry by incorporating synthesis, new types of measurements and experiments. Solid state chemistry had transformed to materials chemistry by assimilating various chemical approaches to the study of materials. These included elements of molecular chemistry as well as of supramolecular chemistry and developments in theory and computation. I guided research in areas such as surface science and computer simulation which were not my primary interests, mainly to arouse interest in the topics and to create expertise. My last student in spectroscopy and molecular structure received his Ph.D degree in 1994. One of my students in spectroscopy (T. Pradeep, now a Professor at IIT Madras) fabricated several electron spectrometers for his Ph.D thesis. The tradition of fabricating instruments has continued in my laboratory. The faculty that I had recruited and the areas we pursued in the new Solid State and Structural Chemistry Unit attempted to cover many important areas of the chemistry of materials. Some of my colleagues have become famous and the Unit has emerged to be a prime contributor to research at IISc.

I received a letter from Purdue University in 1982 informing me of the honoris causa D.Sc degree that the university wanted to confer on me. This was overwhelming, coming from my alma mater in the US. This was my first honorary degree and it was a matter of personal satisfaction to receive it (see the diploma and the citation that follows). As I was recovering from this surprise, I received another message that I had been awarded the solid state chemistry medal by the Royal Society of Chemistry (London). I decided to go to the US to receive the honorary doctorate from Purdue and also planned to stay there for a couple of months to give a special course on solid state chemistry. As I was planning this trip, I received the wonderful news in May 1982, that I had been elected Fellow of the Royal Society, London. G.N. Ramachandran was the first to call me about this. As an FRS himself, he had got advance notice of my election. Being an F.R.S has been considered to be an important scientific recognition in India. A. Ramachandran used to tell me for years that I should get those three letters. My parents were overjoyed.

At the end of my stay in Purdue for two months, I participated in a Gordon Research Conference in solid state chemistry in New Hampshire. Almost all the active practitioners of the subject were there. It was an enjoyable conference. One of my British friends started his lecture at the conference with a joke which went like this. "The Duke of Rochester was dreaming in the House of Lords that he was giving a speech. He then woke up to find that he was actually giving one". The next talk was by another British friend who had come late to the first lecture. He commenced his lecture stating that he would first tell us a joke. He started by saying, "The Duke of Rochester". Everyone laughed.

The quantity and quality of my research were both improving with time and I had a period of intensive research during 1979–83. In 1983, I received an invitation from the University of Cambridge to be the first Nehru professor. I had also been elected professorial fellow of King's College. I recollect an amazing incident that occurred at that time. I had taken my wife for a medical checkup to Patel Chest Institute in Delhi. When I was with her at the hospital, people told me that there

PURDUE UNIVERSITY

UPON THE NOMINATION OF THE PRESIDENT AND THE FACULTY
AND BY THE AUTHORITY OF THE TRUSTEES

HAS ADMITTED

C. N. R. RAO

TO THE DEGREE OF

DOCTOR OF SCIENCE

FOR HIS PIONEERING CONTRIBUTIONS TO MANY
AREAS OF SOLID STATE AND STRUCTURAL CHEMISTRY

WEST LAFAYETTE, INDIANA
MAY 15, 1982

PRESIDENT OF THE TRUSTEES

PRESIDENT OF THE UNIVERSITY

C. N. R. RAO DOCTOR OF SCIENCE

In a letter to the New York Times in 1945, James Bryant Conant, the president of Harvard University, declared, "There is only one proved method of assisting the advancement of pure science — that of picking men of genius, backing them heavily, and leaving them to direct themselves."

Certainly, Conant's observation is all the more true because of the example of Dr. C. N. R. Rao, who is one of chemistry's most prodigious researchers — a man of genius — in the complexities of chemical spectroscopy, molecular structure, and solid-state chemistry.

He is world-renowned and often honored in his field for his impressive contributions to the world's scientific literature which now include 10 books and more than 300 research papers. Most recent of the many prestigious honors that have come to him was his being named a Fellow of the Royal Society in London.

Professor Rao's honors and awards are impressive and include, besides membership in the prestigious Royal Society, the Marlow Medal of the Faraday Society; S. S. Bhatnagar Prize in Chemistry; Sri C. V. Raman Award in Physical Sciences; Centennial Foreign Fellowship of the American Chemical Society; Federation of the Indian Chamber of Commerce and Industry Prize for Research in physical sciences; and the S. N. Boss Medal of the Indian National Science Academy. The president of India bestowed upon him the national honor, Padmashri, in 1974. He was elected a foreign member of the Academy of Sciences of Yugoslavia in 1981.

He is also distinguished by the fact that his wife, Indumati, is an alumna of Purdue, having received the M.S. degree in 1968.

Purdue is proud to count Professor Rao as one of its own.

was a phone call for me from the Prime Minister of India. I could not believe that it could be true. When I took the phone, it was indeed Mrs. Indira Gandhi who congratulated me on being the first Indian to become the Nehru Professor at Cambridge. I had got to know her through my membership of the NCST and of the Science Advisory Committee to the Cabinet. I also used to meet her in connection with the CSIR Society. It was really a wonderful relationship. She knew that I was not looking for any position in Delhi and that I was available to work for the cause of science without any desire for positions and other benefits.

In Cambridge, I carried out research in collaboration with a number of people in electron microscopy, organic solid state chemistry as well as inorganic materials chemistry. John Thomas (earlier in Aberystwyth) was the head of the physical chemistry department. He was a great host and our friendship strengthened through collaboration. I visited Prof. Nevill Mott in the Cavendish Laboratory regularly and we developed a close relationship. Prof. Mott would ask me and my friend Peter Edwards to organize a party in some college. Prof. Mott would come early in the evening to have a drink and stay until late talking to us. He would narrate stories many of which I still remember. A few of the stories belonged to the time of Lord Rutherford. One story that I remember pertained to the time when Nevill Mott started doing his research at the Cavendish Laboratory of which Rutherford was the director. Within a few months, Nevill Mott had made a major contribution wherein he had provided the quantum mechanical proof for Rutherford scattering. Enthusiastically, Nevill Mott went to Rutherford to tell him about the work. When young Mott got there, Rutherford was reading something. He just looked at Mott and said, "Very nice, come back when you do something equally clever". That was the end of the meeting.

Sir Nevill Francis Mott — Rao's Guru

Life in Cambridge was wonderful. We lived in a nice house on Chaucer Road, next to the botanical gardens, which we had rented from the University. Membership of King's College added much to our stay there. The large number of eminent visitors as well as brilliant aspiring scientists bestow great vitality to Cambridge University. I could meet many fine scientists during my stay there, Jean-Marie Lehn (Nobel Laureate in Chemistry) being a notable one. Amongst the many friends that I made in the college, I must mention Martin Rees (now President of the Royal Society), David Husain, Stephen Elliott and Bill Jones (now Head of Chemistry in Cambridge). Fred Sanger, Sydney Brenner and many other distinguished scientists were members of King's. At the college Christmas party, someone came to me and asked "who is that person (without a tie) standing alone in the corner?" I had to tell him that it was Fred Sanger with two Nobel Prizes. When he received the first Nobel Prize, Fred Sanger was offered a knighthood which he had refused. He said apparently, "every one calls me Fred". During our stay in Cambridge, Suchitra and Ganesh stayed with us for a few months. Dr. Ganesh carried out research on nucleic acids in the Medical Research Council laboratories.

A significant accomplishment during my stay in Cambridge was completing the book, *New Directions in Solid State Chemistry* in collaboration with J. Gopalakrishnan. The book, published by Cambridge University Press, has done extremely well and is being used widely all over the world. The book has been published in paperback edition as well, and has been translated into Chinese and Russian.

Recollecting my days in Cambridge (1983–84), I feel that this period was important to my research career, since it gave me an opportunity to widen my horizons not only in terms of the areas of research but also with regard to the way to do research. My association with Prof. Mott and others in Cambridge was specially beneficial. During my stay there, my friend Peter Edwards (now head of the chemistry department in Oxford University) and I planned to edit a Festschrift for Prof. Mott's 80th birthday. This book entitled *Metallic and Nonmetallic States of Matter* was published by Taylor and Francis (1985). I also enjoyed visiting the Royal Society to attend meetings and giving lectures in various parts of Britain.

Towards the end of 1984, I returned to Bangalore. I had just been appointed Director of the Indian Institute of Science. (Ramaseshan who was director during the previous two years had just retired). Satish Dhawan was insistent that I accept the directorship, although I had some reservations. I knew that this was going to affect my research adversely, but had no choice other than to assume the responsibility. I had to right away tackle many serious problems at the Institute, most important of all being to improve the infrastructure on a war footing. The supply of electricity and water was unreliable and insufficient, and depended on wiring and plumbing which were 75 years old. I got help from the Ministry of Human Resources and Development and from the Planning Commission (thanks to Dr. M.G.K. Menon) to accomplish these objectives. After becoming director, I handed over the chairmanship of both the Solid State and Structural Chemistry Unit and the Materials Research Centre to my colleagues. I started each day early in the morning in my laboratory office and went to the director's office around 10.30 a.m. only to go back again around lunch time. I managed to spend at least six to seven hours in my laboratory. Saturdays and Sundays were fully devoted to the laboratory and students. There were difficult problems associated with the support staff. There was a need to improve the budget, recruit good faculty in many of the important areas of science and engineering, build hostels as well as staff housing and usher in essential academic reforms. I had to negotiate with the

With Prof. Ramaseshan

employees on many issues with the help of some of my colleagues. Fortunately for me, the institute council was helpful. Mr. G.K. Chandramani was the Chairman of the council for some time but was soon replaced by my dear friend, Dr. Raja Ramanna.

Soon after I became director of IISc, I visited G.N. Ramachandran who lived close to my house in Malleswaram. I invited him to come back to the Institute and do whatever he wanted. He had left the Institute after retirement and had spent some time in Hyderabad. We could give him an office and an assistant. I was delighted that he rejoined the Institute. In the meantime, he was also conferred a professorship of the academy in Delhi.

1984 was the platinum jubilee year of the Institute. I had requested Mrs. Indira Gandhi to inaugurate the platinum jubilee, but unfortunately she was brutally assassinated in October 1984. Mr. R. Venkataraman, Vice-President of India, kindly agreed to be the chief guest. A history of the institute entitled *Pursuit of Excellence* (written by B.V. Subbarayappa) was released on the occasion. The platinum jubilee helped us not only to recollect the glory of the Institute but also to plan new things. J.R.D. Tata was at his best. A platinum jubilee garden was created over an area of around 12 acres in a corner of the campus. In that process, I decided to convert the entire campus into a beautiful forest and started planting several thousand trees. The trees in the square (in front of the main building) had been planted in the form of a Union Jack when the Institute started. We covered the entire square with trees. The Institute campus can pride itself of having extraordinary plant diversity. A major effort was made to recruit faculty in all the divisions of the Institute with some success.

Soon after the platinum jubilee, I did not feel well. My years of agony and anxiety were probably responsible (at least partly) in clogging my arteries and veins. I visited London with my wife in February 1985 to deliver a lecture at the Royal Society. I went through a check-up in Harley Street Clinic following the advice of my dear friend Dr. K.G. Nair (cardiologist in Bombay). I was advised to immediately undergo bye-pass heart surgery, which I did. (John Wright was the surgeon). I recuperated in Cambridge staying in Robinson College (courtesy of its warden, Lord Lewis). I returned three weeks later to Bangalore via Delhi. In Delhi, I attended the investiture ceremony at the Rashtrapati Bhavan and received the award of Padma Vibhushan from the President of India. I learnt later from Sharada Prasad (Press Advisor to Mrs. Gandhi) that Mrs. Gandhi had nominated me for the award. Also on the cheerful side, Sanjay completed the B.Sc degree from Poona University with a first class in 1985. It was a special year for the entire family. Suchitra gave birth to a son, Kartik, on November 14. He is our only grandson and has been a great source of happiness.

The main quality desired of an administrator in an academic institution is in human management. J.R.D. Tata was known to be outstanding dealing with people. Managing a campus on a day-to-day basis has additional problems. The director is also the mayor of the

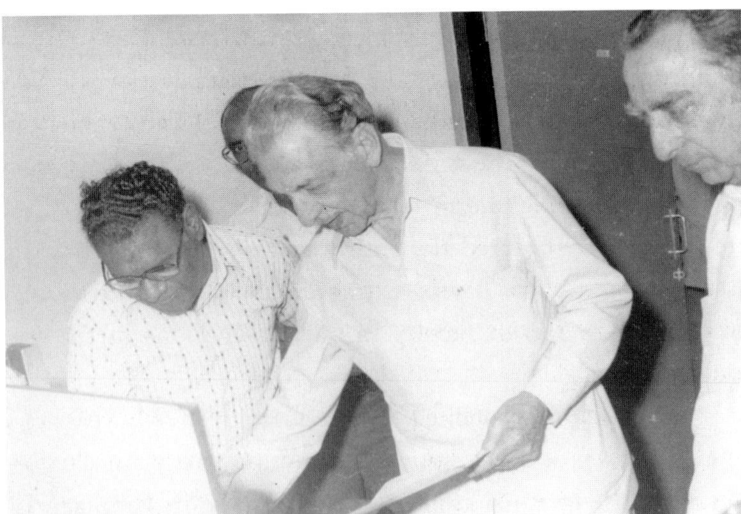

With Mr. J.R.D. Tata, President of the IISc Court (1980)

campus. One of my early accomplishments was to get rid of cattle freely grazing on the campus. The employees who owned the cattle did not like the new arrangement, but it had to be done to save the gardens and the overall atmosphere of the campus. Amongst the several steps taken to help the employees, specially the low-paid ones, were easy loans to buy sites or build houses and grants for medical treatment. When there was a drought in Bangalore, there was acute shortage of water on the campus as well. It was the same time when I was trying to get fresh plumbing done on the campus. One evening, several housewives came to my house with empty buckets to demonstrate. I was just recovering from my surgery.

I had unusual experiences related to the personal problems of members of the supporting staff and faculty. I remember the agony of a housewife who had been thrown out of her house by her husband after marrying another woman. They had lost their only son due to meningitis and the first wife could not have more children since she had undergone tubectomy. The employee wanted another son. I faced several instances where I had to help in settling domestic quarrels in faculty homes. All in all, I had excellent relations with all the members of the Institute. It was easy to deal with good faculty members and the vociferous ones were generally not academically superior. I knew most of the employees by name and it was common for them to wait for me by the road to make some request or the other. This was easy to do since I was a walking director and did not use my car within the campus. It was agonizing to see students suffering due to poor guidance by faculty members. Occasional suicides by students due to unhappy academic experiences were difficult to bear. I must specially mention one case of suicide. The student had completed his Ph.D work and had received satisfactory reports from examiners. On the morning of the final oral examination, he was found dead having swallowed some insecticides. He had left a note stating that he was dissatisfied with the quality of this thesis work (having had poor or no guidance from the supervisor).

The terrible part of the suicide was that the student had already received an offer of a post-doctoral fellowship from America.

As director of IISc, I had some fine people helping me in various aspects of administration. I fondly remember them. I must specially acknowledge the contributions of Mr. K.C. Reddy (retired Secretary of the state PWD) who acted as adviser on civil projects. He ensured that there were no omissions and commissions. With crores of rupees worth of construction going on at the Institute, he saw to it that there were no malpractices. Mr. Chikkappa and other project engineers contributed much to the construction programme. Many of them continued to work with me even after I left IISc. I must remember Professors A. Sreedharan and R. Kumar for their help with employees' problems. Dr. B.S. Subba Rao was an able medical officer who has also taken care of the health of my family as well. I had wonderful private secretaries of whom L.R. Rao and D.V. Seetharaman deserve special mention. I enjoyed working with the gardeners who planted trees and bushes all over the campus keeping pace with my enthusiasm. Debi Sengupta of the Power Engineering Department helped in organizing classical music concerts on the Institute campus. He has been a dear friend with whom I share my fondness for music.

The most annoying habit of some people that is difficult to bear concerns writing of anonymous letters. I used to receive many, generally against the administrators (and a few against me). Curiously, I caught one faculty member sending anonymous letters against a senior colleague who was supposedly his friend. I realized early that one has to learn to ignore anonymous letters and loud mouths. One has to develop a thick skin or better still, become slightly hard of hearing. Sir Harold Thompson at Oxford used to say that it was useful to pretend to be hard of hearing. One can always say, "I am sorry. I do not remember hearing that". I often think of the advice of P.N. Haksar (Principal Secretary to Mrs. Indira Gandhi): "Remember, you will always be assailed by lesser people".

The period 1985–87 was a difficult and busy period. I had just become the vice-president of the International Union of Pure and Applied Chemistry in 1983, to take over as President in 1985. During 1983–85, I was vice-president of the Indian National Science Academy (New Delhi) and succeeded Arun Sharma (botanist from Kolkata) as President in 1985. I had to juggle many things at the same time. There was a conference of scientists and technologists in Delhi in 1985 organized by the Government of India. Prime Minister Rajiv Gandhi inaugurated it and sat through the proceedings for some time. On the afternoon of the inaugural day, I received a message from him asking me to see him that afternoon. We talked about science in India for some time, and he asked me to chair the new Science Advisory Council to the Prime Minister (SAC to PM) that was being set up. I took up this responsibility with his promise that he would take personal interest in the deliberations and recommendations of the Council. Rajiv Gandhi kept his promise.

SAC to PM with non-official members (Roddam Narasimha, Ashok Ganguly, Madhav Gadgil, Jayant Narlikar, P.N. Tandon, R.A. Mashelkar, Shekar Raha, Sam Pitroda and P. Rama Rao were some of

Rajiv Gandhi presenting the G.M. Modi Award to Rao, and next to Rao is M.G.K. Menon

them) was a small, active body which met every month and came out with several important recommendations (more about SAC to PM later). Rajiv Gandhi would see me once in two weeks or so and meet the Council once in 6–8 weeks. My meetings with him were at his house, in his office or in the Parliament. It was a wonderful experience working with an enthusiastic and loving Prime Minister. I remember an occasion when he was to present me an award in his house. There was a huge crowd of citizens waiting to see him in the front lawns. He saw me with my wife in a corner of the lawns and shouted "CNR", walking quickly towards me. Our relationship became even closer after he lost the elections when I would meet him to plan and discuss how government operations should be restructured when he returned to power. Unfortunately that did not happen.

Following the Rajiv Gandhi–Gorbachov declaration in 1985, I was asked to be Chairman of the long-term cooperation agreement between Soviet Union (now Russian Federation) and India. This programme has been used for our benefit in various ways.

In December 1986, there was a major revolution in physical sciences. A high-temperature superconductor had been discovered. The material had broken the long-standing 23K barrier and became superconducting at 35K. I did not know much about it until I met Prof. P.W. Anderson (Nobel Laureate in Physics) who was visiting Bangalore for an international conference on valence fluctuation at the Ashoka Hotel. I gave a plenary lecture at that conference as well, and soon after my lecture, Prof. Anderson asked me whether I knew about this new high-temperature superconductor. When I told him that I had no knowledge of it, he mentioned that the material was an oxide with lanthanum (La), copper (Cu) etc. I asked him whether by any chance, it was related to La_2CuO_4. When he said yes, I told him of the work we had done on this family of oxides many years earlier. My first paper on this family of materials with P. Ganguly had been published in 1971. We immediately went to my laboratory at IISc and spent two hours going through all my

Rao with some of the co-workers who worked on superconductivity

papers. We had already shown that La_2CuO_4 was antiferromagnetic. I started to worry as to how I could contribute to the exciting area of warm superconductors.

I had many sleepless nights, until we discovered two months later, one of the first liquid-nitrogen superconductors, by employing a novel strategy. Till then, all the superconductors required liquid helium. This new compound, $YBa_2Cu_3O_7$ (called 123), characterized independently in Bangalore (at the same time as Bell Labs and Beijing), became superconducting around 90 K (above the liquid nitrogen temperature). We started research on many aspects of high temperature superconductivity, including synthesis of new materials, study of new phenomena and so on. I could obtain support from the Government of India for a national programme on the subject (thanks to Rajiv Gandhi). There were conferences galore everywhere, the first one organized by *Nature* magazine in Boston. They had chosen eight invited speakers including myself. It was an extraordinary occasion for physical sciences, with thousands of people gathering to find out what was happening. High-temperature superconductivity has taken more space in research journals (probably two hundred thousand pages) than any other single discovery in physical sciences. During that time, many a high school student, mixed and ground Y_2O_3, $BaCO_3$ and CuO to prepare the new superconductor. The discovery of high-temperature superconductivity in metal oxides changed the materials scenario in the world, and chemistry gained due importance for its role in the design and synthesis of materials. A recent report of the US National Academy of Sciences nicely reflects on the seminal role of chemistry in materials science.

In 1988, when I was about to visit Switzerland to deliver a plenary lecture at a conference on high-temperature superconductivity, my mother passed away in Bangalore. I cancelled my trip. The tragedy associated with my mother's death is something I can never forget. I was away at a meeting of the Science Advisory Council to the Prime Minister in Goa with my wife. I got a phone call from Bangalore stating that there had been an accident in my house and that my mother was affected. We quickly returned by car from Goa to find that my mother had burnt herself severely while praying in the pooja room. Her saree had caught fire while lighting the oil lamps. She was in a hospital for a few weeks and suffered badly. Indu who was very close to my mother spent every day with her talking to her and helping her to pray. With her passing away, I lost my first guru. I think of her every day. I wonder why such a saintly person had a painful end. When we went through the almirahs and cupboards that she had used, we found nothing. She had given away everything to charity.

When I was once participating in a meeting of SAC to PM in Delhi in 1988, I got a phone call from Mr. Manishankar Iyer, who was the Secretary of the Pandit Nehru Centenary Committee. He informed me that of the various proposals from groups and agencies for commemorating the centenary (1889–1989), the suggestion to set up a new kind of scientific research institute had been approved by the committee. He wanted me to work out a detailed plan. I prepared a programme for the proposed institute with the help of my colleagues in SAC to PM, specially Roddam Narasimha, Madhav Gadgil, R.A. Mashelkar, P. Rama Rao and Jayant Narlikar. One requirement was that the new institute was to be located close to an existing institute of excellence which would provide certain facilities. The plan was soon approved and there was competition as to where the institute should be located. Pune, Allahabad and Kanpur were candidates. All these locations offered land and other infrastructure support. I was asked by some people, why not Bangalore? I immediately approached the Chief Minister of Karnataka (S.R. Bommai) who granted around 18 acres of land for the purpose. Fortunately, Rajiv Gandhi decided that the new institute should be set up in Bangalore and that the Department of Science and Technology would provide the support. I was asked to plan and establish it. I was already the director of the Indian Institute of Science, and had to start building the new institute parallelly. The year was 1989. The new institute was named Jawaharlal Nehru Centre for Advanced Scientific Research (JNCASR). The emblem of the Centre with Nehru's signature was personally approved by Rajiv Gandhi. The famous architect, Charles Correa, planned the first set of buildings for the new campus. We had to have some facilities at the Indian Institute of Science according to the requirement and fortunately, the Chairman of the Council and the trustees agreed to provide space for housing, guest accommodation and an office.

Jawaharlal Nehru Centre for Advanced Scientific Research, Bangalore

Shri. K.R. Narayanan dedicating JNCASR to the Nation

One or two scientists made personal attacks on me at that time, implying that my interest in all this was to become a permanent director or chairman of the Indian Institute of Science and of the new Centre. Another criticism was that in one of the early reports of the Centre, the Academic Coordinator had also included the publications of some of the honorary professors. No one expects a new centre to become famous from papers of others, but the criticism was that the Centre was using the reputation of others to become famous instantaneously. All this was far from the truth. I never wanted to be a permanent director or chairman of any institute or centre. I took up the directorship of the Indian Institute of Science purely to serve and not to get any recognition. In fact, I wanted to give it up as early as possible. Fortunately for me, all my colleagues including Raja Ramanna came to my defence at that time. I also made sure that subsequent reports of the Centre did not list papers of the honorary professors even if their research was supported by JNCASR.

1989 was an important year for materials scientists in the country. A new Materials Research Society of India was created that year. I was the founding president. In 1989, I also completed my five-year term as the Director of IISc. I thought that the time was right to relinquish the directorship, but was urged to continue by Dr. Raja Ramanna, Mr. J.R.D. Tata and the faculty at large. I had to take on another five-year term as director (till July 1994). The post-1989 period was a difficult one. I had started research on fullerenes and carbon nanotubes in addition to my work in other areas of solid state chemistry. (Fullerenes were made in the laboratory for the first time in 1990 and carbon nanotubes were characterized in 1991.) I was forced to sacrifice quality to some extent and missed working on some good ideas due to my administrative responsibilities. It did not mean that I had no time for research. The problem was with the quality of time. When employees' problems, faculty complaints, new Institute programmes and such issues occupy one's mind, it is difficult to carry out high quality research which requires unperturbed attention, and time to meditate. The only satisfaction I derived was that I did my best for the Institute which I loved. I wanted IISc to become more competitive and famous as a post-graduate research institute. This meant closing down some courses and starting new ones. One of the innovations was the integrated Ph.D programme in science. Several new units were initiated in biological sciences and other areas. Computer Science was given a boost and a Society for Innovation set up. The budget of the Institute increased several-fold and excitement about research was evident in the academic community.

One of the saddest events in 1989 in our family was the passing away of Narayanaswamy, my beloved father-in-law. We got a call from my mother-in-law that he was not feeling well. We rushed to Jayanagar where they lived, with an ambulance. When we reached there, he had already died. Narayanaswamy was the most popular person in the Bapu family. He was a fine civil engineer who had built dams (Linganmakki dam in Jog being one of them) and townships. There were some similarities between him and my mother. Both were orthodox and kind-hearted. The unforgettable qualities of Narayanaswamy were his good nature and sense of humour. A typical joke he narrated was: What is the difference between Westerners and Indians, he would ask; the answer was, Westerners dress for dinner and Indians undress for dinner. Narayanaswamy is missed terribly by all who knew him. My wife and I have established a prize in his name for the best Master's thesis in JNCASR.

I was the President of the Indian Academy of Sciences, Bangalore, during 1989–91. It was an enjoyable and useful office to hold where one could serve the wider scientific community. In this capacity, I was an ex-officio member of the council of the Raman Research Institute. A very sticky problem arose at that time which unfortunately affected close relationships and long-standing friendships for some time. The Raman Research Institute wanted to ask a senior professor who had served for some years to leave the institute. All the members of the council were of the same view (including Dhawan and Ramaseshan). But I had to take a strong stand against the move since it was against all known academic traditions and unethical in terms of human considerations. My stand was not dictated by the popular public opinion at that time and I made it clear to Prof. M.G.K. Menon who chaired the council meeting. Fortunately, the person was not removed from the institute, but he voluntarily left the place after about a year (as he had promised me).

1990 was academically significant. I was elected a member of the US National Academy of Sciences (NAS) as well as of the Pontifical Academy of Sciences. Election to NAS is considered to be a great professional recognition. I remember Dr. Kelkar telling me that I should not only get into the Royal Society, but also to NAS. I was already a member of the American Academy of Arts and Science, but NAS is different. It is a difficult academy to get into. A number of extraordinary scientists in the US whom I know are not members of NAS and some got elected only after they received the Nobel Prize. I was elected a foreign member of the American Philosophical Society (APS) in 1995. APS is the original American academy started by Benjamin Franklin. A friend of mine from the US wrote to me at that time, "You have done the hat-trick, becoming a member of all the three science academies of the US". Election to APS is considered to be a special recognition and the letter of invitation to the membership is itself interesting (see the letter on the next page).

During 1990, I worked with Mr. Chandra Shekar's government as member (science) of the planning commission. It was a short-term experience (see Chapter 5 for details) which made it clear to me that my place was in the laboratory, something that I knew all the time.

Kelkar passed away in 1990. He left the world as quietly as he had lived. His passing away did not make big news in the country, but he had left behind a great tradition in institution building and a large group of admirers. Prof. D.K. Banerjee passed away in 1993. He was a health fad and I expected him to live till the age of 100. He used to visit me periodically in the Institute and I would offer him a cup of hot milk to drink as he had given up drinking tea. In memory of this close friend and great organic chemist, my wife and I have donated some money to the organic chemistry department at IISc for a lecture to be delivered by a young, up-coming organic chemist.

As luck would have it, another major development occurred in 1993–1994 in the chemistry and physics of solids. The phenomenon of colossal magnetoresistance (CMR) was found to occur in a family of metal oxides, the rare earth manganates. Besides CMR, these materials exhibit other novel properties, specially electronic phase separation. The presence of two or more distinct phases with different electronic properties in the same crystal is truly interesting and new to physics. I had worked on these manganates earlier, and initiated a major programme on the new phenomena in late 1993. I could do this satisfactorily since I had given up the directorship of IISc on July 31, 1994. Dr. Raja Ramanna and several others wanted me to continue for

AMERICAN PHILOSOPHICAL SOCIETY

HELD AT PHILADELPHIA FOR PROMOTING USEFUL KNOWLEDGE

104 South Fifth Street, Philadelphia, PA 19106-3387

Phone: (215) 440-3400 • FAX (215) 440-3436

Herman H. Goldstine — Executive Officer

April 21, 1995

Dr. C. N. R. Rao
Indian Institute of Science
Solidstate and Structural Chemistry Unit
Bangalore 560012
INDIA

Dear Dr. Rao:

This notification of your election today as a foreign member of the American Philosophical Society, on a form used by the Society for a very long time, may require an explanation.

The American Philosophical Society was founded by Benjamin Franklin in 1743, in emulation of the Royal Society and the Dublin Society of Arts. The APS has played an important role in promoting science and learning in the United States and recognizing scholarly achievement everywhere. Lavosier, Darwin, Einstein, Pasteur, Lord Rutherford, and Mme. Curie have preceded you among our foreign members.

Under separate cover you will receive a copy of the 1993 Year Book of the Society, which contains the Laws and List of Members, a short history, and a full account of the Society's recent activities. The 1994 edition will not be published until this summer.

As you will see, today the Society supports research through a program of grants, publishes scholarly books and journals, maintains a library singularly rich in material in the history of science and technology. Its meetings express in the twentieth century the universal spirit of the eighteenth century "Age of Reason" with the presentation of papers on such topics as Jefferson and the ambiguities of liberty, superconductivity, cancer control, the history of computers, rising health care costs, Shakespeare, and education and our divided knowledge.

The next Autumn General Meeting will be held on November 2 - 3, 1995. The next Annual General Meeting occurs on April 25-27, 1996. Both meetings occur in the newly renovated Benjamin Franklin Hall. The building opened in 1993 when the Society celebrated the 250th Anniversary of its founding and of the birth of its third President, Thomas Jefferson. Meanwhile, after we receive your acceptance of election, your certificate of membership will be sent to you.

Sincerely,

separate cover

Philosophical Hall • 1789 Benjamin Franklin Hall • 1855 Library Hall • 1959 another five-year term. Fortunately, I could convince them that the time was right for me to give up the directorship. I went to the University of Wales for a two-month period before relinquishing the directorship so that my successor G. Padmanaban, a distinguished molecular biologist, could familiarise with the office. For the first time, I was at peace with myself after July 1994. My 60th year was celebrated by my students and colleagues with an international conference. Four leading journals in the world published special issues on that occasion. World Scientific brought out a volume containing my selected papers on solid state chemistry. Before giving up the directorship, I got a vision document prepared along with a master plan of the campus. It was most gratifying that the council of the Institute passed a resolution that allowed me to take research students as long as I wanted, and also conferred the honorary fellowship of the Institute on me. My only hope was that I left the Institute in a better shape than before. The physical plant of the Institute had certainly improved several-fold.

Indu retired as reader in education from the teachers' college in 1995. She also gave up her responsibilities in the school which she had started. She decided to take up educational technology as her prime occupation and learnt all that she could about multimedia. She has since been working in an honorary capacity in the education technology unit of JNCASR. Sanjay also learnt various aspects of computers including multimedia and has been working in this unit. In collaboration with Jatinder Kaur and Sanjay, Indu has produced excellent materials for science education. The CDs from this group are highly useful and instructive to students and teachers. Using these CDs, multimedia presentations are being made all over the country. These presentations generally accompany an hour-long lecture by me (power point presentation) on "Learning Science", "Celebration of Chemistry" or "Nanoworld". These road shows have been presented to thousands of children in various locations in the country. Typical of the locations are Kolkata, Mumbai, Delhi, Bangalore, Chennai, Siliguri, Gauhati,

Bhavanagar, Palampur, Trivandrum, Tezpur, Dharwad and Sikkim. Such direct contact programmes with school children, specially in rural areas, are of crucial importance for the future of science in the country.

A great event in our family was the arrival of our granddaughter Suguna in December 1995. Suchitra gave birth to her in Pune. Suguna stole my heart the minute I saw her. She continues to dominate my feelings.

I was not involved in any major committees in the country after July 1994. The only commitment I had was with the Indo-Japan Science Council which I had helped to establish in order to promote closer relationship between Indian and Japanese scientists. I visited Cardiff for a few weeks as an honorary professor in 1993 and 1994. Wyn Roberts was my host in Cardiff and my old friend Brian Smith (later Sir) from the Berkeley days was the vice-chancellor. In 1995, I was invited to be a Distinguished Visiting Professor in the Department of Materials by the University of California, Santa Barbara (UCSB). I was in the School of Engineering at UCSB with Venky Narayanamurthy (later dean at Harvard) as the dean. The Materials Department at UCSB is fantastic with outstanding scientists as its members, several of whom are members of the U.S. National Academies and three are Nobel laureates. Matt Tirell who succeeded Venky as the dean was himself a fine scientist. My dear friend Tony Cheetham was the director of the Materials Research Laboratory at the university. Harry Kroto (Nobel Laureate and discoverer of fullerenes) was the other distinguished visiting professor. I have spent around four to five weeks in the summer months in Santa Barbara for 14 years during 1994–2007. It was during this period that I started research on open-framework structures and hybrid materials. Collaboration with Tony Cheetham was helpful in this respect and we have published several papers together. I was also able to contribute some new ideas to this area. The work on colossal magneto-resistance and related phenomena exhibited by manganates flourished in Bangalore in a big way.

In 1995, Prof. Nevill Mott reached the age of 90. Peter Edwards and I edited a book on metal-nonmetal transitions in his honour (Taylor and Francis, 1995). Peter Edwards and I had organized a discussion meeting on metal-insulator transitions in the Royal Society in 1996. Prof. Mott had expressed a desire to participate in it. Unfortunately, he passed away at the age of 91 before the meeting. I had lost a great guru.

When Mr. Deve Gowda became the Prime Minister of India in 1997, I was asked to chair the Scientific Advisory Committee to the Cabinet. I accepted this assignment but before I could take up the chairmanship, Mr. Deve Gowda was replaced by Mr. I.K. Gujral as Prime Minister. Mr. Gujral encouraged me to take up the chairmanship and it was a pleasure to work for him. He was a fine prime minister in the tradition of Pandit Nehru. Unfortunately, his government lasted for less than a year, but even within that short period we could make some useful contributions. During this period, I helped to initiate a new professional society to promote chemical research. The Chemical Research Society of India has been welcomed by the chemists of the country, and meetings of the society have been invigorating.

Despite the many activities, I had to undergo a re-do heart surgery in early 1998 in Breach Candy Hospital in Bombay. There was again a problem with the plumbing and I accepted the advice of Dr. K.G. Nair. Dr. Panda was the surgeon. In September 1998, my wife and I visited France where I was to deliver plenary lectures at two international conferences. Just as I completed my first lecture in Versailles, a secretary came running to inform me that my father had passed away in Bangalore. He was 94 years old and was slowly losing control of his limbs. He had suddenly lost vision in one eye which had devastated him since he could not read newspapers easily. He used to take long walks in the mornings till his 92nd year.

I have participated actively in the programmes of the Third World Academy of Sciences (TWAS), now called the Academy of Sciences for the Developing World, since the late 1980s. I am a founding member of TWAS and became vice-president and later president of this Academy. Towards the end of 1992, I assumed responsibility for the election of new fellows to the Academy and for the selection of TWAS prize winners. I tried to improve the quality of the fellowship and brought some norms for the election of Fellows. TWAS has provided the only forum for scientists of the developing world and has emerged to become a powerful academy representing the aspirations of the least developed countries (more about TWAS later).

The Royal Society awarded me the Hughes medal for physical sciences in 2000. The medal was in recognition of the work on transition metal oxides. This medal had been awarded to Prof. C.V. Raman a few decades ago. In 2003, the government of the state of Karnataka bestowed on me its highest honour, Karnataka Ratna. There was a special function in the civic theatre where the Chief Minister (S.M. Krishna) presented the award (I am delighted that Bhimsen Joshi, the great musician, was given this title in 2008). 2003 was also a sad year. Sivaraj Ramaseshan passed away after a brief illness. Satish Dhawan had passed away earlier in 2002 most unexpectedly. I had lost my three close friends from the early IISc days. Satish was the moral conscience of the academic community in India.

I turned 70 in June 2004, and 70 of my selected research papers were published as a volume by World Scientific. There was an international conference on solid state chemistry in Bangalore to mark the occasion. A few journals brought out special issues for the occasion. There were a number of friends and colleagues to felicitate me. It was nice to see Justice Nittoor Srinivasa Rao at the function. He was nearly 100 years old. I spent some time talking to him about the old days. He remembered my grandparents' house in Basavanagudi. He even recollected the name of the person who lived next door to my grandparents.

Dr. Manmohan Singh became Prime Minister of India in 2004. I received the first India Science award by the Government of India in that year. Dr. Manmohan Singh asked me to be Chairman of Science

Advisory Council to the Prime Minister (SAC to PM). It has been a pleasure to work in this Council to usher in new initiatives in the science and technology scenario of the country. Investment in science and technology and higher education has also increased enormously in the last 2–3 years (more about this SAC to PM later).

Without waiting for the campus to come up, JNCASR started some activities as early as 1990. Typical of the activities were summer research fellowships for undergraduates, discussion meetings and a visiting scientist programme, all from an office in the IISc campus. Professors V. Krishnan and N. Mukunda were of great help in organizing these programmes. The Centre started research and other academic activities on the new campus only in 1994. Faculty recruitment and admission of research students were initiated, and the Centre has since grown. JNCASR was inaugurated by Mr. K.R. Narayanan (then Vice-President of India, and later President) in March 1995. There are departments dealing with materials, molecular biology and genetics, evolutionary biology, engineering mechanics and new chemistry. The Centre confers Ph.D degrees as well as M.S. degrees by research. The integrated Ph.D programme in materials science is one of its kind. Today, the Centre has around 180 Ph.D and M.S students and entertains a large number of visiting scientists and student interns. It offers several innovative programmes in science education and supports the development of useful educational materials. The Centre has international collaboration agreements with leading institutions in the world.

Building JNCASR has been a labour of love. It is really a pleasure walking through the beautiful campus of JNCASR, adorned by a well-planned garden. I was the founding president of JNCASR during 1994–1999. I gave up the presidentship in 1999 and the council of management was kind enough to confer the honorary fellowship of the Centre, and to name me honorary president. Reliance Industries endowed a large sum to the Centre to establish the Linus Pauling Professorship. (Thanks to the efforts of Ramesh Mashelkar and Manmohan Sharma.) This endowment will be treated as a donation to the Centre when I relinquish the professorship. V. Krishnan (a fine inorganic chemist) succeeded me as President, and Prof. M.R.S. Rao (a distinguished biologist) has been the President for the last six years. JNCASR has been served by many loyal and capable officers. Nagaraja Rao and A.N. Jayachandra have been administrative officers of the Centre. For over 15 years, I have had an outstanding personal secretary, Shashi Karthikeyan. She has excelled others in this capacity through her loyalty and single-mindedness.

I became interested in the synthesis and properties of small clusters or nano particles of metals in the 1980s. An intriguing question at that time was, "How many atoms make a metal?" This problem actually arose out of a question that I was asked by an undergraduate student. It took me a few years to find out that one needs around 300 atoms of gold to render it metallic. Such particles of gold would have diameters between 1 and 2 nanometers. Metals become nonmetallic when the particle size is reduced sufficiently. Gold does not shine when the particle size is less than 2 nm. I have been working on nanoparticles, nanowires, nanotubes and a variety of nanostructures for some years. A noteworthy aspect of the contributions made from Bangalore was the work on inorganic nanotubes as well as on new varieties of carbon nanotube structures. Another novel contribution has been the use of the oil-water (organic-aqueous) interface to generate nanocrystalline films of inorganic materials. Nanoscience has become highly popular in the last 10 years and I have enjoyed working in this area immensely. A national nanomission has been initiated by the Government of India with me as its chair.

What is interesting about research in advanced materials is that quite often new and exciting materials are discovered and research strengthens in new directions. This happened again in the year 2004 with graphene. Graphene is nothing but a single sheet made of six-membered rings of carbon atoms. The sheet, only 0.34 nm thick, exhibits most unexpected properties and phenomena. Research on graphene has made me younger in the last three years.

I have been collaborating with physicists for over four decades. This collaboration got intensified after the high-temperature superconductivity season. V.G. Bhide, T.V. Ramakrishnan, N. Kumar, Ajay Sood, Arup Raychaudhuri and Milan Sanyal are some of the physicists who have collaborated with me extensively. I have enjoyed working with theoretical physicists and chemists without whom many of the experimental results from my laboratory could not be understood. In recent years, Umesh Waghmare and Swapan Pati have been of great help in this regard.

Over the years, I have received honorary doctorate degrees from many universities in India and abroad, the first one being from Purdue in 1982. Of the various ceremonies, I specially recollect the encaenia in Oxford which was conducted in Latin in the Sheldonian in contrast to the one in Northwestern which was in a football stadium. In Colorado, it was in the basketball stadium. In Uppsala, a canon is fired as the rector places a wreath of leaves on the recipient's head and hands over a

Rao at Purdue University

Desikottama award at Shantiniketan

Signing the register at the Sheldonian theatre, Oxford

Honoris Causa at Northwestern University, USA

gold ring with the recipient's name on it. The royalty sits in the audience in the front row and joins for tea later. In Shantiniketan, the Desikottama title was awarded in the house of Rabindranath Tagore, some of the proceedings being in Sanskrit. The Nadoja title was awarded in Hampi on the banks of Tungabadhra, with the proceedings conducted in Kannada.

May 2008–May 2009 was the centenary year of the Indian Institute of Science. Several programmes were conducted by the Institute to celebrate the centenary under the leadership of its director, P. Balaram. It was a pleasure to be associated with them. In particular, I enjoyed delivering the inaugural centenary lecture and participating in the inaugural function. I specially remembered C.V. Raman, J.C. Ghosh and Satish Dhawan on this occasion. A book containing 60 of my research papers (entitled *Trends in the Chemistry of Materials*) was published by World Scientific as a centenary publication of the Institute. There was a function with Abdul Kalam (former President of India) as the chief guest where I was asked to say a few words. Dr. Manmohan

Singh inaugurated the Institute centenary in December 2008. During this visit to Bangalore, he also dedicated to the nation, the new International Centre for Materials Science (ICMS) as well as the Hall of Science in JNCASR. ICMS has been created to promote high-quality research and international collaboration in the area of materials science. The Hall of Science has been established to conduct science education programmes for students and teachers through the year, and to produce novel educational material such as CDs and books. The new building for the Hall of Science has been constructed with the donations from the C.N.R. Rao Education Foundation and the Reliance Industries Ltd. In the Hall of Science, we conduct direct-contact programmes for students and teachers on various science subjects. A chemistry laboratory has been set up where school students carry out experiments using a kit developed by Dr. S.N. Bhat, who directs the laboratory. He also oversees

Opening of the CNR Rao Hall of Science and dedication of ICMS to the Nation by Prime Minister Dr. Manmohan Singh (December 3, 2008)

the project-oriented chemical education programme for undergraduate students from the colleges of the country.

The 33-year period between 1976 and 2009 has been a phenomenal period for my academic pursuits. I have learnt much and carried out research in various areas of interest. The seeds of solid state and materials chemistry were sown in the early 1960s. I could, however, make noticeable accomplishments only by the end of the 20th century. It is only around the year 2000 that I began to feel that my students and I may have accomplished something meaningful. This does not mean that we have accomplished everything that we wanted to or that we had reached the pinnacle of research. Not by any means. My students and I have published many research papers and we are getting cited extensively. I have been elected to almost all the major science academies of the world. I shared the one-million dollar Dan David Prize for materials science (considered to be the Israeli Nobel) in 2004 with George Whitesides (Harvard) and Robert Langer (MIT). Science is not, however, about getting awards or writing papers. It is a way of life.* It provides unlimited opportunities to contemplate and to accomplish in an area of one's choice. It involves climbing the limitless ladder of excellence. It has been wonderful to keep climbing this ladder. As I continue to climb this ladder, I have also become 75 years old (June 30, 2009). Four international journals and two Indian journals brought out special issues to mark the occasion. In 2009, the Royal Society awarded me the Royal Medal (Queen's medal). Considering that the Royal Society had already awarded me the Hughes medal for physical sciences, this was specially overwhelming. My 75th year coincided with the 75th year of the science academies of India and of the centenary of IISc. I had the privilege of delivering the inaugural lectures of the diamond jubilees of the national science academies in Delhi and Bangalore.

I felt truly flattered by the recent recognition from Germany through the award of the August-Wilhelm-von-Hofmann Medal for outstanding

* See the essay on "Doing science as a way of life", presented as the Epilogue.

contributions to chemistry. This medal (initiated in 1902) is given once in two years to a leader in chemistry in the world. I feel honoured to be in the list of recipients of this medal.*

As I am growing old, I see my contemporaries in the family and in the world at large disappearing. Except for our immediate family, my wife and I are alone after my mother-in-law passed away in 2007 at the age of 91. More recently, we have lost dear cousins. Some of my past students are no more, but life goes on, and fresh batches of students keep my spirits up. It has been a pleasure to see our grandson Kartik grow up to become a doctor. He is pursuing higher studies in medicine. He is also our consultant on books and films.

On old age, it is amusing to remember the famous anecdote involving Winston Churchill. When he reached his 80th year, a 25 year-old artist was commissioned to paint his portrait. After he completed the task, the young artist turned to Churchill and said, "Sir Winston, it has been an honour to paint your portrait. I do hope that I will be able to do this again when you turn 90". Churchill looked at the young man and said "Why not! You look quite healthy". I could cheerfully end this narration of my long journey in chemistry with this anecdote, but I would not be honest if I did not make a confession on the feeling of disappointment and dissatisfaction that I constantly experience. I am reminded in my conscious (and subconscious) moments that the goal is farther away and it may never be reached. Curiously, I often think of J.C. Bose (the first modern scientist of India) in search of the origin of the Ganges river. He walked day after day until one day in desperation he asked someone passing by, "Where is the source of Ganga?" The passer by showed him the peaks of Himalayas far away that had to be crossed before reaching the source of the Ganges. Bose started walking towards the distant spot. When he eventually reached it, he fell unconscious.

* It was a matter of great honour that the Royal Society asked me to deliver in the presence of Queen Elizabeth II and other members of the royalty, the Response on behalf of the scientists during the 350th Anniversary in London.

Receiving Padma Vibhushan from the late President, Zail Singh (1985)

Pope John Paul II confers the Fellowship of the Pontifical Academy of Science and presents the medallion (1990)

Prime Minister Dr. Singh presenting Rao the India Science Prize (2004)

Rao receiving the 'Chevalier de La Legion d'Honneur' award from France (2005)

Rao receiving The Royal Medal at The Royal Society (2009)

Rao receiving the 'Abdus Salam Medal for Science & Technology' (2009) from Jacob Palis, President of TWAS

Rao being conferred the 'Order of Friendship' by the Russian Deputy Prime Minister (2009)

THE · FARADAY · SOCIETY
THE MARLOW MEDAL
ALBERT EINSTEIN
UNESCO
भारत विज्ञान पुरस्कार
सत्यमेव जयते
भारत सरकार
सी.एन.आर. राव
2004
INDIA SCIENCE AWARD
सत्यमेव जयते
Government of India
C.N.R. RAO
2004
RAMAN CENTENARY MEDAL · IACS CALCUTTA NOV 1988
DAN DAVID PRIZE
Professor C.N.R. Rao

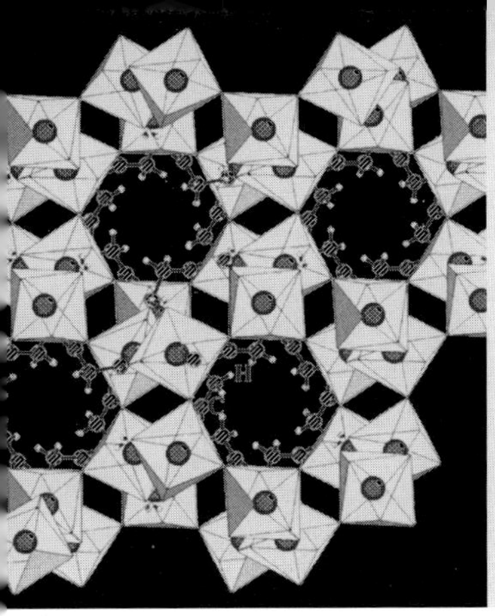

5

Labyrinths of Power

When I was young, I was always hesitant to meet administrators, commissioners, secretaries to government and so on. When minor officials from a ministry came to our institute, people looked up to them and treated them like demi-gods. When heads of institutions behaved in such manner, the young people would get more nervous. This situation probably still exists. Administrators have somehow built such an aura about themselves that there is a feeling in most of us that without them the country cannot move forward. There may be some truth in this statement, since civil servants do run the country independent of political situations.

Power — is it mythical or what exactly is power? People who run institutions and governments are considered powerful. In what way? Power in governments appears to manifest itself in different ways — in controlling the way things take shape at the rate desired by administrators or authorities. It manifests itself in denying something that is due. It would not have manifested itself as worthy of mention if everyone did his/her job, followed established principles and procedures, and carried out decisions faithfully, avoiding delays. Power comes into play because individuals become conscious of their positions and do things their own way. This happens more frequently than not through negative actions. Power is often the consequence of negative actions. I have heard from ministers that many of their decisions and recommendations in the

Parliament do not see the light of day, because bureaucrats do not facilitate matters. Many industrialists and technologists state that it is easier to work with politicians than with bureaucrats. This peculiar beast called bureaucracy is acutely felt in developing countries. It is only when bureaucracy becomes less important, can one consider a country to be developed. We can see why India is definitely a developing country. Unfortunately, power breeds corruption, and the hunger for power spreads to all walks of life from ministers to the lowly policeman.

Some of the administrators develop a cynical or a patronizing attitude. It is not unusual to witness some of the scientists who take up administration transforming into bureaucrats. They soon forget their roots, yet wanting to be treated as eminent scientists. It is difficult to talk to them since they seem to know everything already. I have known young administrators who are helpful and cheerful, but as they go up in rank, they stop conversing. They lose their sense of hearing and talk down to ordinary mortals. Having said all this, I must also mention that there are administrators who are not so conscious of their positions and actually help to solve problems. Many a great administrator is self-effacing, with a positive attitude.

I remember a joke made by a friend of mine who used to head an important agency in the government. "A person comes to complain about a problem. I send him away saying that I will look into it. A year later, the person comes back and thanks me for having got things done. I would have actually done nothing. The file would have somehow moved on its own."

A serious limitation in the administration of government is the short tenure of the secretaries. For example, the department of non-conventional energy in the Government of India has had at least 15 secretaries in the last 20 years, in spite of the importance of non-conventional energy to the nation.

It is difficult to blame bureaucrats and politicians alone. Our society gives undue importance to them and many scientists prefer politicians to inaugurate scientific symposia. There are some meetings where little happens after the chief guest goes away. The opposite is the situation in Sweden where ministers take public transportation to come to meetings and sit with the rest of the participants. Even in a scientific committee or commission, the secretaries get more importance than the scientists. It is probably the importance that bureaucrats enjoy in society that makes administration attractive to some of the young people and scientists. I have known people who have taken positions of secretaries to government, giving up positions in industry, academia or elsewhere.

I remember my first experience of a government meeting. I was a young professor at IIT Kanpur and was asked to chair a meeting in the ministry of petroleum and chemicals. When I went there, I found that the secretary of the committee was a senior civil servant, an ICS (who had worked for decades with the British rulers). I was shy and found it somewhat embarrassing to deal with a senior secretary to government who was expected to produce the minutes of the meeting. It took me some time to get used to chairing committees.

In 1969, when I was 35 years old, I was asked to become chairman of a public sector undertaking (Hindustan Insecticides (HIL) Ltd). HIL had factories in Delhi, Cochin and Maharashtra. I used to chair the meeting of the Board of Directors and shareholders and so on. It was a useful experience to deal with a company and its tribulations. With the help of a fine managing director, we set up research laboratories of the company and opened new ventures. The company even started making profit. What was interesting, however, was that the joint secretary in the ministry determined many of the policies. I must admit that my association with HIL and later with IPCL (Baroda) as well as the Reserve Bank of India gave me valuable experience, although I decided to get out of such commitments several years ago to conserve time for research.

I got a glimpse of the powers at work in the government when I was appointed a member of the National Committee on Science and

Technology (NCST) in 1971 with Mr. C. Subramaniam (CS) as the Chairman. CS was truly outstanding. People would make all kinds of presentations to him, but he would eventually come out with the right decision and the right conclusion. CS did not want heads of agencies to be members of the committee. NCST was to have only non-official members. This had advantages and disadvantages. The disadvantage was that secretaries of science departments in the government were generally not happy with the council although a large body of scientists and technologists was consulted in the various deliberations. The advantage was that we could discuss issues without the overbearing influence of administrators.

In retrospect, an excellent science and technology plan was indeed brought out by NCST for the very first time in the country. One of the first recommendations of the committee even before it started preparing the S & T plan was to create the Department of Science & Technology (DST). This met with opposition by some of the heads of agencies. The Director-General of CSIR commented "why do we need a new department for science and technology? What can it do that CSIR cannot?" Similar reactions were noted from other quarters. Fortunately, Mrs. Gandhi could be convinced that there was need for such a department to coordinate important S & T programmes, to fund scientific research and to take new initiatives.

A major effort of NCST was to integrate S & T into socio-economic planning. Each member of the NCST had a portfolio. My portfolios were scientific research and higher education, and advanced materials. In the panel related to scientific research and higher education, Raja Ramanna, Arcot Ramachandran and Rais Ahmed were some of the members. In the panel on materials, E.C. Subbarao, S. Ramaseshan, Abdul Kalam and D. Chakravorty were some of the members. With the help of the advisory panels, we came out with important recommendations which were discussed widely with the scientific community at large. One important recommendation was to set up a Science and Engineering

Research Council (SERC) to fund research in the country and Regional Instrumentation Laboratories to provide analytical and other services to scientists in universities and other laboratories. SERC has served the scientific community in a big way for the last 35 years or so. One of our recommendations was to establish four national science universities in the country, but this did not materialize. In the area of materials, we prepared a report on the status of advanced materials and a futuristic plan for the area.

As a member of NCST, I had some interesting experiences. I had convinced Dr. Nayudamma who was the director general of CSIR the need for a dedicated national centre for advanced materials research. He wanted me to write a description of such a centre and asked me whether it could be created within the existing CSIR institutions. I suggested that it could possibly be created on the campus of the National Physical Laboratory. There would be a close relationship between the Centre and the NPL, with the NPL director being administratively responsible for it. The Centre itself could be autonomous. I wrote the description of the Centre and Dr. Nayudamma liked it. Eventually it was forwarded to the director of NPL. After a few days, the NPL director told me "Dr. Rao, I can give you a deputy director's post similar to many other senior people here, but not much more than that". I was shocked. I was not looking for a job. The academic merit of the proposal was completely lost. The Centre was not established.

There were also some pleasant surprises. I was asked to chair a committee on science museums and submit the report to the planning commission. Being young and inexperienced, I advocated the setting up of a National Council of Science Museums which would organize, coordinate and manage a chain of science museums in the country. The committee supported the idea enthusiastically. After submitting the report, I completely forgot about it. To my surprise, sometime later the National Council of Science Museums was established, and it is thriving.

During the NCST days (1971–74), we met Mrs. Gandhi many times to discuss the progress made by us. During the period of the Indo-Pakistan war, we would meet her to discuss the scientific needs of war efforts. I remember at least one occasion when we met around 10 p.m. in her house. My contact with Mrs. Indira Gandhi increased with time and I would see her at other meetings such as the CSIR Society as well. I also witnessed ingratiating or meek behavior displayed by some scientists. I remember a particular scientist who would vociferously praise Mrs. Gandhi during the meetings. It was embarrassing to watch.

As a member of a national commission, I was involved in the appointment of the secretary of the commission to replace the person who had just retired from service. We discussed possible candidates, including the seniormost person in the organization. We went away after the meeting. During the next meeting of the commission, the chairperson started the meeting stating, "As we had agreed in the last meeting, we should now approve the appointment of a particular person as the secretary". I objected to what was happening. I told the commission and the chairperson that we had not discussed the appointment of any individual and had only looked at some of the possible candidates. There was no way that I would go along with the appointment of a predetermined candidate. I asked every member individually whether we had agreed on any particular person for appointment in the previous meeting. The members would not say yes or no. They were all embarrassed. I soon offered to resign from the commission. The proposal to appoint the person concerned was withdrawn.

I was a member of the Scientific Advisory Committee to the Cabinet (SACC) during 1979–83. Prof. M.S. Swaminathan was the chairman and Prof. M.G.K. Menon was Member-Secretary of SACC. After some months, Swaminathan left the country to join the Rice Research Institute in Phillippines and Menon took over the Chairmanship (Manju Sharma acted as Secretary of the Committee and later became Secretary of DBT). This SACC accomplished many things. For example, the Department of Biotechnology was established at that time based on the recommendation of the committee. COSIST (Committee to Support Infrastructure in Science and Technology) was established by the UGC at that time. M.M. Sharma and I worked together on this initiative. Some of the efforts did not succeed. One of them was to make the director-general of ICMR an important person in government since its director-general was generally ruled by officials of the health ministry. ICMR did not have sufficient autonomy and was poorly funded. Mrs. Gandhi could be convinced about offering the director-general of the ICMR the status of a secretary to the government. Unfortunately, this was not allowed to happen by the bureaucrats.

I experienced the ambience of the corridors of power more intimately, when I was Chairman of the Science Advisory Council to the Prime Minister Rajiv Gandhi (1985–89). The Council with only non-official members, had P.J. Lavakare of the DST as its secretary. Rajiv Gandhi had kept the science portfolio to himself and would have meetings with the Council in the Parliament house, at his residence or in his office.

With Rajiv Gandhi and members of the Science Advisory Council to the Prime Minister (1988)

Mr. K.R. Narayanan (later President of India) who was minister of state for science and technology attended the meetings. Rajiv used to call me frequently to discuss important issues. Quite often, ministers would be waiting to see the Prime Minister. Initially, I felt awkward that he was talking to me for an extended period when the finance minister or someone else was waiting outside. I remember an occasion when I was with him late in the evening explaining the terrible situation faced by scientists in importing chemicals, spare parts, etc. and the duty imposed on such imports. He immediately asked the finance minister to come to his office and told him to solve the matter that very evening, sitting with me for as long as it takes.

Our Council handled many issues at that time, one of them being the establishment of the Technology Information Forecasting and Assessment Council (TIFAC), based on an initial paper prepared by M.G.K. Menon (who was Science Adviser to the Government at that time besides being Member of the Planning Commission). The setting up of the technology development fund was another novel idea. Reviewing the seven technology missions for literacy, oil seeds, communications and so on was a fruitful experience. The Council recommended that Sam Pitroda be asked to be in-charge of the technology missions. The establishment of the Centre for Developing Advanced Computing (CDAC) in Pune and many other such efforts became possible because of the recommendations or intervention of the Council. Rajiv Gandhi referred several important items to the Council for advice, typical ones being the Antarctic expedition and the need of the cryogenic engine by the space department for advanced rockets. The Council deliberated on such issues with the help of experts and made its recommendations. When the report of the CSIR review committee was submitted, many difficulties were identified. Scientists in the CSIR were perturbed by some of the recommendations. Rajiv Gandhi asked SAC to PM to go through the recommendations and come out with modifications, if and where necessary. Such tasks kept SAC to PM busy.

We made important suggestions which included the setting up of a central advisory board for CSIR as well as technology advisory boards for different areas (they were set up by the then director-general, and were disbanded subsequently).

SAC to PM brought out reports on important areas related to science and technology. A somewhat serious limitation that we faced arose from the absence of whole-hearted cooperation from the secretary of the Department of Science and Technology. He was upset that he was not a member of SAC to PM, but there was nothing that we could do about the matter. Rajiv was emphatic that the Council should have only non-official members in their personal capacity.

I found Rajiv to be a wonderful person to work with. He was kind and generous. He gave extraordinary importance to science and technology through his personal interest. He always asked intelligent and penetrating questions. Those who were not prepared found him difficult and this was common amongst secretaries and agency heads in the government.

I worked closely with a number of officials in connection with the work of SAC to PM. I remember my close relation with Montek Ahluwalia who was then in the P.M.'s Office. Sarla Grewal and B.G. Deshmukh who were both Principal Secretaries to the P.M and Cabinet Secretaries, one after the other, were helpful.

An interesting incident during those SAC to PM days, relates to the appointment of the head of an agency. I was a member of the selection committee and the minister chaired the meeting. Raja Ramanna was sitting next to me in the committee. After some discussion, the committee decided to select a particular person as the head of the agency, but the minutes were not written. At the end of the meeting, I asked the minister about the minutes. He said, "Whatever you have decided will be in the minutes". It was amazing that a few days later, when the appointment of the agency head was announced, the person was different from the one we had chosen.

SAC to PM also deliberated on the science and technology interface with higher education and made some suggestions to be incorporated in the national policy for education. I would meet Mr. P.V. Narasimha Rao who was the education minister to discuss important issues. He generally agreed to the suggestions made by SAC to PM. We very much wanted to see that a National Council for Science and Technology was established with the Prime Minister as the Chair in order to provide a forum to discuss broader issues and to determine priorities. This could not be done. Such a Council would also ensure dependable support for science and technology, independent of the political party in power. (M.G.K. Menon tried to establish such a council when he was a minister in the V.P. Singh government, but unfortunately the government did not last long enough.) Another recommendation that did not see the light of the day pertains to exploitation of solar energy. We had set a target of 1000 MW of solar photovoltaic power by the year 2000. We did not even reach 10 MW.

Rajiv took considerable trouble to be present for important meetings and occasions. I remember how he came to Kolkata just for the evening to be present for the Raman centenary. He was there on time to inaugurate the Indian Science Congress in Pune (1988) of which I was president. He was misunderstood by some. Let me give one instance. There was a presentation of a national mission by a secretary to government. During the presentation, the secretary could not read what was on the transparency being projected on the screen. Rajiv told him that he expected a better presentation. Some people immediately commented on how Rajiv wanted fancy presentations. This was not true. I always used overheads with my hand written transparencies. I had no problem whatsoever. There was another occasion where in the organizational chart for a big mission, the concerned secretary to government had shown a lower division clerk as a requirement (in a separate box). How could anyone keep silent by such lack of sense!

Towards the end of the term of SAC to PM, Rajiv Gandhi asked us to work out how we could increase the investment on science and technology from 1% to 2 or 3%. If that money became available, how would it be spent? When we were deliberating on these issues, the general elections were announced. During one of my meetings with Rajiv in the Parliament house, a heated debate was raging on the Bofors issue. Some members were caustic and abusive. I asked Rajiv why he was not responding with a strong reply. He said, "CNR, no one will believe me now, whatever I may say. One thing, I can tell you. If I or a member of my family is found guilty of any misdemeanor, we will accept any punishment that is accorded". When I thanked Rajiv for sparing some time during such a hectic and harrowing period, he replied, "In political life, there will be something important or catastrophic everyday. There is no point waiting for a free day".

The way some politicians and bureaucrats behaved after Rajiv Gandhi's term as Prime Minister was somewhat frightening. There was an oppressive atmosphere where everyone was made to feel nervous because of the possible guilt one may or may not have carried. Every item had to be screened for corrupt practices. Any major equipment to be purchased from abroad was to be justified for no bribes and middlemen. An officer had been appointed specially for the purpose. IISc was planning to buy a supercomputer from the US at that time and Dr. Raja Ramanna asked me to see this officer. I remember how nervous I was when entering his office. I do not know why I had to be afraid.

The Government works in complicated ways. Some occurrences may be without its full knowledge. Some others do not happen even with the full support of the head of the ministry. A joint secretary or someone below can torpedo a project. Many tasks do not get done since secretaries are shifted often. There are ways senior bureaucrats are punished or pushed aside. They are sent as advisers (often to the planning commission) with no one ready to listen to their advice!

When Mr. Chandra Shekar became Prime Minister in 1990 with the support of the Congress party, I was in Trieste for a conference. I received a midnight call from Rajiv Gandhi who asked me not to be perturbed by what I would soon hear. Rajiv had asked the new PM to take my advice in matters related to science and technology, but Mr. Chandra Shekar had announced my name as Member (Science) of the Planning Commission, a position I had never wanted. Rajiv asked me to meet him soon after I returned to India and not to take any decision before that. When I met him, he asked me to work for some time (a few months at least) as Member of the Planning Commission and contribute to the plan, specially the S & T part. Mr. Mohan Dharia was the Deputy Chairman of the Planning Commission. I knew him well when he was the Minister of State for Planning during the NCST days. I was allowed to be in Bangalore as director of IISc and partake in the work of the Planning Commission. It was my feeling that the Planning Commission was much too bureaucratic (and a bit of an white elephant). As Pandit Nehru had said, what was to be a think tank involving great minds had transformed to a huge establishment with dark labyrinths.

My association with the Planning Commission was short-lived (1990–91), since Mr. Chandra Shekar resigned a few months later, but I did gather some experience listening to presentations of the chief ministers of various states and working with economists in deciding priorities. Within that short period, we had prepared a plan for science and technology which was liked by Mr. Mohan Dharia and others. Mr. A.N. Verma who was Secretary of the Planning Commission became a good friend. Mr. Chandra Shekar was easy to deal with. He was straight-forward and supportive. I had visited him in his home a couple of times (once with Dr. Manmohan Singh who was the economic adviser). We had to sit on the floor on carpets in his ashram-like office. I was with Mr. Chandra Shekar on the day he resigned. We had a meeting of the CSIR Society in the Parliament house. Mr. Chandra Shekar asked me to chair the meeting on his behalf and went away. He came back after an hour and a half, by which time the agenda had been completed. I reported to him on what was discussed, and he concluded the meeting. He had just then returned from the Rashtrapati Bhavan after submitting his resignation. I could see lots of politicians roaming in the corridors. General elections took place soon after.

Rajiv Gandhi was assassinated by mad terrorists during an election campaign. I can never forget that day. I could not control my agony. He was an idealist and a great friend of science. He would have made a great Prime Minister in his second term. He was the last Prime Minister who personally took care of the science and technology portfolio. In Rajiv's memory, I organize the annual Rajiv Gandhi science and technology lecture delivered by an internationally acclaimed scientist. The lecture is supported by the Rajiv Gandhi Foundation and Mrs. Sonia Gandhi chairs these lectures. George Porter, Charles Townes, Harry Kroto, John-Marie Lehn, Werner Arber, Michael Atiyah, Ahmed Zewail, Roald Hoffmann, Martin Rees, David Gross, Alan Heeger and Dick Zare are some of the people who have delivered this lecture.

When the government of Mr. P.V. Narasimha Rao came to power, I expressed my desire not to continue as member of the Planning Commission. Mr. A.N. Verma became principal secretary to the Prime Minister. I got a phone call from Mr. Narasimha Rao's office to see him within a few days after he assumed office. I met him in his house early one morning, and made a short presentation on the state of science and technology and of the various efforts made during the previous years to integrate science and technology with socio-economic planning. He seemed happy with the presentation and asked me rightaway to take up the chairmanship of the new Science Advisory Council to the Prime Minister that he was going to appoint. He asked me to suggest a panel of names for the membership within the next two days. I did so. In the meantime, Mr. Rangarajan Kumaramangalam, who was the Minister of State for Science and Technology, submitted a proposal that there should be a vice-chairman of SAC to PM, who should be paid full salary (and suggested the name of a person). I was against this because all of us worked in an honorary capacity. This problem postponed the formation of the Science Advisory Council indefinitely. There was no science advisory mechanism during Mr. Narasimha Rao's tenure. It was not a great time for science and technology in the country. I wrote and spoke about the sad state of science at that time in several places, including the science academies. This angered Mr. Narasimha Rao who would not talk to me after that. Towards the end of his term (when Kumaramangalam was no longer in the cabinet), I was asked whether a SAC to PM could be formed. I saw no point in doing so when there was little time to do anything meaningful.

I was chairman of the Science Advisory Committee to the Union Cabinet (SACC) when Mr. Gujral was the Prime Minister (1997–98). We could do one or two things then. The Principal Secretary to the Prime Minister was Mr. Vora who was very helpful and efficient. He told me that there was no need for me to come to his office to see him on matters to be executed. "You just make a phone call or send me a note. I will take care of the rest". In fact, this is how we established FIST (Fund for Infrastructure in Science and Technology) for supporting universities. It was done by first receiving the approval of the Prime Minister and then sending a note to Mr. Vora.

During my years of exposure to the powers in the government, Cabinet Secretaries and Principal Secretaries to Prime Ministers have generally been cooperative. There were one or two occasions when I had to be irreverent to some of the top bureaucrats. This became necessary whenever I felt that the concerned person was dismissive and condescending. Many secretaries to government helped me immensely when I was director of IISc in matters related to budgets, employee benefits and so on. Secretaries can also be quite difficult in decision making. Let me recollect one instance. Mr. J.R.D. Tata wanted to establish the National Institute of Advanced Studies (NIAS) on the campus of IISc. This required the approval of the trustees of IISc. The secretaries of education and law of the government were both opposed to the idea of giving IISc land to the new institute being set up by the Tatas. Mr. J.R.D. Tata then talked to Mr. Rajiv Gandhi who then had a chat with me. The land was given to NIAS after a few weeks with the agreement of the members of the IISc trust. Raja Ramanna was the first director of NIAS. I served as an acting director for close to a year when Ramanna went to Delhi to serve in the V.P. Singh government.

When the NDA government came to power (1999–2004), Dr. Murli Manohar Joshi was the minister of science and technology. (The NDA government came to power after the Gujral government.) I informed him that the SACC still had a term of a year and a half left and would be happy to work with the new government. He had no interest in SACC. The NDA government did not appoint any science advisory committee. I was slowly eased out of all the committees of the government. Some of the committees had just started functioning (with me as chairman), but I was no longer involved with any of them. Dr. Murli Manohar Joshi had a meeting of scientists to draft a new national technology policy statement. I was not invited to that meeting. It was hurtful then, but it was actually a great blessing too. I started fully devoting my time to research and my work took a turn for the better. The quality of my publications improved substantially and citations of my research papers increased close to 3000 or so per year. There was a great increase in my impact index as well. I must thank the NDA government for the great favour it did by eliminating me from all the committees.

My tenure as the Chairman of the Science Advisory Council to the Prime Minister (SAC to PM) during the period of Dr. Manmohan Singh (2004–2009) has been productive and enjoyable. This SAC to PM had only non-officials as members, with secretaries of science departments and heads of agencies as observers. Roddam Narasimha, Ashok Jhunjunwala, Bikash Sinha, P. Rama Rao, Syed Hasnain, Kiran Karnik, Swati Piramal, K. Vijayaraghavan, M.M. Sharma, Sanjay

Dhande, Baldev Raj, P. Balaram, G. Mehta and T.V. Ramakrishnan were some of the members. Secretary of DST was member secretary V.S. Ramamurthy and later T. Ramasami have held this position. Ramasami is a chemist and has been a friend of many years. Mr. Kapil Sibal was minister of science and technology and I could work with him easily. He became a personal friend. SAC-PM came out with several recommendations, many of which were approved by the government or have been executed. Two of the important decisions taken by the government based on the recommendations of the Council are the establishment of a new department for earth sciences in the ministry of science and technology and the creation of a department of health research, the secretary of which would be the director general of ICMR. The latter was accomplished 30 years after it was first mooted during the time of Mrs. Indira Gandhi. The setting up of the five new Indian Institutes of Science Education and Research (IISERs) has been a major

With Prime Minister Dr. Manmohan Singh and Shri. Kapil Sibal at the release of the SAC-PM report

contribution of the Council. These institutes are now functioning in Kolkata, Pune, Mohali, Bhopal and Trivandrum. I was chairman of the board of IISER, Kolkata for the first three years (with S. Dattagupta as the Director) and continue as chair of the Standing Committee of the government for IISERs.

The national nano mission has supported research and education in nanoscience and nanotechnology in a big way. Several nano centres and two nano institutes have been set up to provide an impetus to this important area. A new National Science and Engineering Research Board, similar to the US National Science Foundation, was approved by the Parliament. This board will be responsible for taking the necessary steps to support scientific research suitably and adequately, and also improve the quality of research. It will bring out annual national science indicators. The creation of new types of fellowships for recognizing talented scientists at various levels (such as the J.C. Bose Fellowship and the Ramanujan Fellowship) has been welcomed by the scientific community. There are many other proposals and recommendations of the Council which are in the process of being executed. The Council also made recommendations to help research scholars and scientists, in terms of service conditions, emoluments, and so on. More has been accomplished during the term of this Science Advisory Council to the PM than any equivalent period. Dr. Manmohan Singh met the Council on many occasions to discuss the recommendations and the meetings were very useful. I would meet him personally in his house to appraise him of the various deliberations and to discuss new initiatives of the Council. These meetings were always friendly and encouraging. It is gratifying that Dr. Manmohan Singh has decided to increase the investment in S & T to 2% of the GDP from the present 1%. I trust that this will be achieved during the second term of Dr. Singh as Prime Minister. He has asked me to continue as Chair of SAC-PM for another term. The new council retains some members of the previous one and has a few younger members in addition. I must mention here that I have willingly devoted my time and effort for science in India, but have avoided being in government administration or being a member of the upper house of the Parliament.

I worked closely with the Ministry of Human Resources and Development in 2005–2009. The minister (Mr. Arjun Singh) and the secretary (late Mr. Sudeep Banerjee) were highly cooperative in establishing the new IISERs and in several other initiatives. With the cooperation of the University Grants Commission, an empowered committee was set up to provide additional grants and other forms of support to universities. I was chairman of the board of governors of IIT Kanpur for a term (Sanjay Dhande was director), but more significantly, I have chaired the standing committee of the Council of IITs for sometime. Very few academic issues were, however, brought up for discussion in the standing committee by the directors of the institutes. I was taken by surprise when eight new IITs were suddenly announced in 2008 and students were admitted without proper discussion in the standing committee or elsewhere. There was neither a campus nor a director for any of the new IITs, even a year after the students were admitted. I discussed this matter with the Prime Minister and a number of other responsible people, but have not found the reason why it had to be done this way. We have to protect IITs in order to ensure that they remain as Centres of Excellence. IITs should be great centres of innovation. Some weaknesses of these institutions in this respect were noticed long ago by the first review committee of IITs of which I was a member and Nayudamma was chair. If action is not taken, they will gradually become glorified engineering colleges.

There are some issues that we were unable to settle during the 2004–2009 term of SAC to PM. We have tried to improve the participation of women in science, also the conditions of workplace and employment for women scientists. We have had partial success at best, mainly because most men do not really accept this as an important issue. We could not minimize bureaucratic procedures in the approval of appointments, promotions and extensions of senior scientists. It takes months to get government approval for the appointment of directors of institutions and this has hampered the functioning of educational and research institutions. We could not identify the rules of government that would encourage mobility of scientists. It is also important to see that once budgets are decided, expenditure on research and development and related efforts is handled by the concerned institution or department without too much of interference and control. It is often difficult to get funds for the second or third year of a project because of the procedural delays. Unfortunately, scientific institutions continue to be treated like any other government set-up such as a district office.

Although I have lived in Bangalore for many years, my association with the Government of Karnataka has been limited. When Mr. Devaraj Urs was Chief Minister, he had asked me to be a member of a committee to review universities in the state with Dr. K.N. Raj as Chairman. We took a lot of trouble to write a good report. By the time, we completed the report, there was another chief minister who asked us to meet him in his house. When we gave him the report, he told me in Kannada, "Tell me, why are we fussing about universities so much? Look at me. I don't have a university degree, but am the chief minister". I was later chairman of a committee which drew up a report on the open university for the state. Early in 2009, Mr. Yeddyurappa, the present chief minister asked me to head a vision group on science. We have prepared an action plan for the state and some of the programmes have been initiated.

I have tried to provide a glimpse of my experiences of the corridors of power. The main purpose of writing this section is because power and bureaucracy have affected all walks of life and have determined the pace of progress of the country. I have avoided details and not mentioned my association with many important institutions such as Tata Institute of Fundamental Research and organizations such as the Atomic Energy Commission (AEC) and the Reserve Bank of India. I have worked closely with successive chairmen of the AEC, specially Dr. Anil Kakodkar. Kakodkar is a fine person and is one of the best AEC chairmen we have had. Of the many science secretaries that I have worked with, I must mention M.K. Bhan of the Department of Biotechnology with whom I share many ideas and ideals. K. Kasturirangan has been a successful chair of the space commission and a close friend.

Although we seem to be making progress on many fronts, I have the feeling that bureaucracy in India has increased with time. I hope that the situation will change in the near future. Many of us are still afraid of bureaucrats and centres of power. When will we lose this complex? Unless there is a change, it will be difficult for science and technology to thrive in India, and for India to become a serious player on the global scene. An India devoid of bureaucracy and corruption can indeed make quantum leaps in science and technology as well as in various sectors of development.

6

Great Fraternity

One of the most pleasant aspects of being a scientist is the opportunity to belong to a large fraternity. One gets to know people from everywhere through science, giving a sense of belonging to the world at large. Best friends may not necessarily be from one's own country. This is true in my case. It is not because I have participated in many international conferences or because I have been a professor in a few places around the world, but due to collaboration in research and my involvement in several international scientific organizations. Such association has made me realise the bigger role of science and also of the greater power of science.

My exposure to the bigger world of science started when I became a member of the chemical education committee of the International Union of Pure and Applied Chemistry (IUPAC) in 1971. My association with IUPAC continued till 1990. Within three years after I became member of the chemical education committee, I was asked to be chairman of the committee. Alexandra Kornhauser (Ljubljana) and David Waddington (York University in U.K) worked with me closely those days. We held chemical education conferences all over the world. Notable chemical educators like George Hammond, Marjorie Gardner, Art Campbell, George Pimentel and Morris Sugden participated in the meetings. I specially remember the meetings in Poland, Slovenia, Ireland, Washington D.C and Madrid. Membership (and later chairmanship) of the spectroscopy commission of IUPAC brought me close to a number of spectroscopists in the world, specially those in Japan, Europe and America. International spectroscopy conferences were frequent during 1960–90.

I was elected a member of the Bureau of IUPAC in 1979, thanks to the nomination by the Indian National Committee. I was next nominated for the vice-presidentship of IUPAC in 1983. It was a difficult election because a well-known professor from Germany had also contested for the position. I got elected because of the support of a number of countries, specially US, UK and Russia. The meetings of the Bureau and the executive committee of IUPAC took me to many places in the world. After two years of vice-presidentship, I became president of the Union for the period 1985–87. It was enjoyable meeting people from all over the chemical world. IUPAC dealt mainly with nomenclature of elements and compounds, atomic weights, reference data on spectroscopy, thermodynamics and so on. The main accomplishment that I can claim as president of IUPAC was the initiation of efforts to restructure IUPAC in order that it reflected modern trends in chemistry. IUPAC gave me an opportunity to work with many great chemists. Some of the close friends from those days are Ted Becker (my old collaborator from NIH, Bethesda), Mostafa El-Sayed (Georgia Tech), Joshua Jortner (Tel Aviv), Saburo Nagakura (Japan) and Valentin Koptyug (Novosibirsk). I made a departure from the traditional presidential address in 1987, with an experimental demonstration of how a magnet is repelled by a high-temperature cuprate superconductor (123) at liquid nitrogen temperature. During my association with IUPAC, I was also involved with the executive committee of the Committee on Data for Science and Technology (CODATA) of the International Council of Scientific Unions (ICSU).

Even before I became director of the Indian Institute of Science (IISc) in 1984, when I was in Cambridge, a fine Swedish gentleman, Mr. Ostholm, came to see me. He had the vision of establishing a research centre of the Swedish Astra company in Bangalore which would work on diseases such as malaria. He needed the cooperation and collaboration of IISc. It seemed a good idea and I tried to be of help when I returned to Bangalore. IISc gave some space to start research on malaria and helped in various ways to establish the Astra Research Centre. I was a member of the board of which Sune Bergstrom (Nobel Laureate) was chairman. Sune and I became good friends as he used to visit Bangalore a few times a year. As soon as he came to Bangalore, he would call me at home and spend an hour or so talking about various matters, having tea with us. Many Swedish scientists came to Bangalore to work in the Astra labs. J. Ramachandran, my old pal from Berkeley, took over as the head of the Centre and established it in new premises not far from IISc. Today, the Centre is much bigger and is part of Astra-Zeneca.

I was a member of the first Indian science delegation to China led by M.G.K. Menon. This was just after the cultural revolution. We visited many laboratories all over China and found them to be in a poor state. They were poorly lit and the corridors were full of cabbages and other vegetables meant for distribution to the workers in the laboratories. We were wearing top coats in the laboratories to bear the November cold. I took a scientific delegation to China two or three years later in 1985 from the Indian National Science Academy. The situation was not different. Look at China today. What a change in a short time! What is curious, however, is the near absence of collaboration between Indian and Chinese scientists and Academies.

One of the important objectives of the International Council of Scientific Unions (ICSU) in the 1970s and 80s was to fight for the rights of scientists to freely participate in international scientific meetings, independent of the political leanings of countries. We faced a problem when Prof. M.G.K. Menon was nominated by India for the Presidentship of ICSU. India had refused visas to scientists from South Africa to participate in an international conference. ICSU took strong objection to this stand and there was some effort to take action against India. I was then President of the Indian National Science Academy and had to argue the case for India. My argument was that the apartheid policy of South Africa was more immoral and inhuman than anything else, and one should, therefore, respect the morality of the Indian stand. ICSU could not block Prof. Menon's election and he defeated the Swiss candidate with overwhelming majority.

I was elected to the Bureau of ICSU in the early 1990s and my association with it continued till 2000. This was a different type of experience where broader issues of science were involved. Typical of such areas were energy, climate, earth system science and science education. I soon became a member of the executive committee of ICSU (Werner Arber, Nobel Laureate, was President). The meetings of the ICSU bureau and board generally took place in Paris. I fondly remember the endless walks and discussions in cafes with Jacob Palis, a well-known mathematician from Brazil. He has been one of my dear friends since. I almost became Vice-President of ICSU in 2000, but pulled out of the election, because I got heavily involved with the Third World Academy of Sciences (TWAS) around that time.

During my association with IUPAC and ICSU, I was also involved in certain international programmes of UNESCO and other international organizations, specially in science education. By and large, I found UNESCO programmes to be poorly funded, considering the lofty objectives. During some of the IUPAC and ICSU meetings, I had the opportunity to deliver lectures to science teachers. On one occasion, I gave a lecture on chemistry while Leon Lederman (Nobel Laureate and particle physicist) gave one on physics. It was an enjoyable event with each of us doing one's best to make the presentation attractive and enjoyable. At the end of the programme, Leon suggested that we should do the Lederman–Rao road show all over the world. I remember an interesting joke by Leon at the beginning of his lecture. "A well-known scientist had been asked to give a lecture to the inmates of a high security

Receiving the UNESCO Albert Einstein Gold Medal from Dr. Federico Mayor, Director-General (1996)

prison. The audience was full of murderers, rapists and convicts. He did not know how to start the lecture. He could not say, nice to see you. Neither could he say, nice to be here. He then decided to say, I am glad that you are all here". In the 1990s, a new International Organization of Chemistry for Development was formed with Prof. Glenn Seaborg (Nobel Laureate from Berkeley) as the Chairman. I have been a member of the council of this organization which created many new bonds with people.*

The Science Institutes Group (SIG), headed by Phillip Griffiths (Institute of Advanced Study, Princeton), has been active in setting up institutes of excellence (Millenium institutes and the like) in developing countries. Brazil, Chile and Mexico have benefitted from this programme. Unfortunately, there has not yet been success with Bangladesh and Vietnam. I have been a member of SIG from the beginning along with Jacob Palis (Brazil), Tom Hexner (Boston) and Harold Varmus (Nobel Laureate in Medicine). Varmus is promoting the idea of a science corps (similar to the peace corps), but it is yet to take shape. My

* Nobel Laureate J-M. Lehn is now the Chairman.

With Prof. Jacob Palis and Prof. Mohamed H.A. Hassan

With Prof. Anthony Cheetham

friendship with Jacob Palis has been enduring over the years, not only through our work in SIG and ICSU, but also because of our common ideals. His generosity and loyalty are truly remarkable. It was during a meeting of the advisory committee of the Brazilian millennium institutes that the terrible 9/11 catastrophy occurred in New York. I was in my room in a hotel on Copa Cabana (in Rio de Janeiro) during a short break at the meeting. I switched on the TV to see the news on CNN. As I was watching, CNN showed a plane going through one of the towers of the World Trade Centre. The commentator (Brown) first said that it was a terrible accident. As I was watching, another plane hit the second tower. What an experience!

In the last two years, I have been participating in an international programme to make chemistry popular and to increase curiosity amongst children in a programme called Molecular Frontiers, led by Dick Zare (Stanford) and Bengt Norden (Sweden). I am also working with Tony Cheetham to help establish a materials research centre in Ras al Khaimah (UAE) with the enthusiastic support of Sheikh Saud.

At a personal level, my connections with the world of science got strengthened by my election to the membership of a number of science academies. The first major academy that I got elected to was the Royal Society, London, in the year 1982. The Royal Society is one of the oldest science academies of the world. I was in Paris on my way to the US to give a lecture at an American Chemical Society meeting and saw my name announced in the Times of London. This was a wonderful piece of news. I have participated in several symposia and meetings of the Royal Society.

The US National Academy of Sciences (NAS), of which I have been a foreign associate for some years, is one of the finest science academies. I have been able to attend very few meetings of the NAS, but I recognize the considerable influence it wields on science in the US and the world at large. A memorable occasion is when I got elected as a member of the Soviet Academy of Sciences. I had gone to Russia for a discussion of the Indo–Soviet scientific programmes, and the meeting was in Kishinev in Moldovia. It was during that meeting that my election to the membership of the Academy was announced. There was a big celebration with champagne flowing. The Soviet Academy of Sciences was a powerful body in the USSR. It does not exist any more and has been replaced by the Russian Academy of Sciences of which I am a member.

Academician Marchuk (Former Deputy Prime Minister & President of USSR Academy) and I were co-chairmen of the Indo–Russian long-range programme in science and technology for several years since 1986. Unfortunately, due to poor health, he gave up the chairmanship two years ago. For many years, my closest friend in Russia was Valentin Koptyug, who headed the Siberian branch of the USSR Science Academy. Koptyug visited India many times and we worked together in IUPAC as well. Unfortunately, Koptyug died a few years ago. In the Soviet days, I used to sometimes feel nervous while visiting Moscow and other places since everything was organized by the officialdom. I would be sometimes worried whether someone would pick me up to take me to the airport or elsewhere. However, no such thing happened. My wife and I fondly remember our ten-day visit to various places in the USSR (including Siberia and Uzbekistan) in the early Gorbachov period.

During the Soviet days, I visited many countries of the old communist block, specially Hungary and Poland. I participated in several conferences in Poland, and one of my dearest friends there is Henryk Ratajczak (Former Rector, Wroclaw University) with whom I have collaborated on problems related to molecular interactions. Our common friend was the late Orville-Thomas (University of Salford, U.K) with whom we have written papers and edited a book. Yugoslavia was another country that I have visited several times. It was one of the few countries where Indians did not require a visa. I had close associates in Ljubljana (now in Slovenia).

The membership of the Pontifical Academy of Sciences (PAS) has been an interesting one. I have attended several meetings at the Vatican and enjoyed walking in the gardens and spending hours in St. Peter's Basilica. The plenary session of PAS generally has discussions on the origin of life or some aspect of science and spirituality. PAS has around 65 members from all over the world and is the oldest science academy in the world, going back to the time of Galileo. I have met many fine people during PAS meetings and got to know some of them closely, Charles Townes (Nobel Laureate in physics), Werner Arber (Nobel Laureate in biology) being two such. The present chancellor of PAS, Bishop Sorondo, is a dear friend. During the plenary sessions of PAS, the Pope gives special audience to the members. Based on the recommendations of the PAS, the Vatican exonerated Galileo in 1994. This was done by Pope John Paul in a public meeting attended by ambassadors of many countries and other dignitaries. My wife and I were both in the audience on this momentous occasion.

Of the many talks that I have attended in the PAS, I will never forget the one by Dr. Joseph Murray. Dr. Murray was a young surgeon during the Second World War. He had to carry out operations on a pilot who had come out of a burning aircraft. The pilot had no facial features left. Murray promised the pilot that he will give him a face and a body that he need not be shy of. He accomplished this after innumerable surgeries. Murray did similar things with many patients. Murray's involvement with patients in distress, with disabilities and deformities, made him look for a way to do kidney transplants. His battle with the problem of transplanting organs is beautifully described in the book, *Surgery of the Soul*. Murray received the Nobel Prize for successfully carrying out kidney transplantation. It has been a pleasure to know him personally.

Induction of a member to the French Academy in Paris is carried out with great gusto and the member concerned stands under the cupole of the Academy (located on the bank opposite Louvre). My association with the French Academy has been interesting, but has not helped in my collaboration with French scientists. My collaboration with French scientists started much earlier. As I mentioned before, Paul Hagenmuller of Bordeaux is a professional contemporary of mine in solid state chemistry (although he is at least 10 years older). We have visited each other several times. Paul is a jolly person and his life has been a colourful one. He was a prisoner in a German concentration camp and could build his school of solid state chemistry only after the Second World War. He was in Vietnam before returning to France to start his professional career. In the last few years, Paul has been restricted to an old age home after undergoing serious surgery. Jean Rouxel (the first student of Hagenmuller) built a great institute of materials chemistry in Nantes. This fine friend died unexpectedly in the late 1990s due to cerebral hemorrage. I have known Bernard Raveau (Caen) for a few decades. We have collaborated extensively and have authored two books. Another friend of mine, Gerard Ferey, has been the head of the solid state chemistry laboratory in Versailles. I had collaboration with Rousset's group in Toulouse and several young people from there have come to work with me in Bangalore. My links with Grenoble started through my friendship with Benoy Chakraverty and a visiting professorship at the Joseph Fourier University. I use the synchrotron in the Institute Laue Langevin (ILL) at Grenoble extensively.

I do not want to go on describing my association with the science academies any further, except to mention the fine lecture tour of Japan that my wife and I had after I was elected to the Japan Academy as an honorary member. This academy has only 60 or 70 members and a special feature of this Academy is the periodic presentation it makes of the status of Japan to the emperor. Another feature is that the average age of the members is very high (probably 80 or more). It is no wonder that two of the Japanese chemists that I know became members after they received the Nobel prize.

I have visited Israel a few times and consider Jerusalem to be one of the greatest cities of the world where there is confluence of three great religions. I have walked for hours in old Jerusalem. I have worshipped at the holy wall and spent time in the environs of the Temple Mount and the al-Aqsa mosque. I have walked along via Rosa remembering the torturous journey of Jesus to the crucifix. My wife enjoyed Jerusalem when we went together to Israel to receive the Dan David Prize.

The Max Planck Institute for Solid State Physics in Stuttgart, Germany, is an outstanding centre of research in areas of interest to me. I have lectured there a few times and some of its directors such as Manuel Cardona, Arndt Simon, Manfred Rühle and Martin Jansen have become good friends. Martin Jansen is especially close to me and has visited Bangalore many times. Another great German chemist who is an intimate friend is Herbert Roesky of Gottingen (where I was Gauss Professor). Achim Mueller of Bielefeld is a friend whom I have known for decades, first as a spectroscopist and later as an inorganic chemist. I am really fond of Peter Gölitz, Editor, *Angewandte Chemie* and its sister journals, whom I consider to be the greatest editor of chemical journals.

Visiting professorships in foreign universities have enabled me to know many scientists better and to collaborate closely with some of them. Visiting Professorships at Purdue, Oxford, Cambridge and UC Santa Barbara have been specially useful and productive. I still collaborate with some of my friends in Oxford and Cambridge. In Purdue, I collaborated with George Honig between 1958 and 2005. With Tony Cheetham (now in Cambridge), I have been publishing continuously for the last several years. My regular visits to UC Santa Barbara got me involved with the International Centre for Materials Research of which Tony was the director. I was the chair of the advisory board of this Centre. This Centre has organized workshops and schools on materials science all over the world (Berlin, Oxford, Trieste, Singapore, Bangkok, etc). The workshop in Vietnam (Halong Bay) was specially memorable. I respect the Vietnamese for their determination. They are also very hospitable. Tony and I have organized international schools and workshops on materials science in JNCASR, Bangalore, with support from UC Santa Barbara and Cambridge University. With an association of over 30 years, Tony has become part of our family. My frequent visits to UC Santa Barbara has resulted in many friends of whom I must mention three great physicists — Jim Langer, David Gross and Walter Kohn. The last two are Nobel laureates in physics. I consider Walter to be a great statesman of science and a fearless crusader on issues such as nuclear disarmament and solar energy. Walter is a gentle person who made his career in Canada and later in the US, after escaping from a Nazi concentration camp when he was a child.

When I was Linnett Visiting Professor and fellow of Sidney Sussex College in Cambridge, I got to know David King closely. He was the head of the chemistry department and later became science advisor to Tony Blair, the British Prime Minister. Peter Edwards, my collaborator of many years, now heads the inorganic chemistry laboratory at Oxford. My chemistry friends whom I have got to know through conferences, memberships of academies and so on are large in number. Some of the young people whom I have met when they were students in Oxford, Cambridge and elsewhere have become leaders in their subjects. I keep in touch with many of them.

The Indo–Japan Science Council was established about 16 years ago with the hope that scientists of the two countries can come closer. The first co-chairman on the Japanese side was Saburo Nagakura whom I have known for many years. We had a common guru in Prof. Robert Mulliken (Nobel Laureate) of the University of Chicago. Both of us worked on charge-transfer spectra and hydrogen bonding, and have participated in conferences together. That old friendship has continued and we still meet each other. Prof. Nagakura is over 90 years old and is one of the most respected scientists of Japan. He does not fail to meet me whenever I am in Japan.

The Indo–Brazil Science Council has promoted cooperation in science between the two countries in the last 4–5 years. My knowledge of Brazil, however, is much older. I have visited Brazil to deliver lectures in various places and the Brazilian Academy of Sciences has been my host during some of the visits. Jacob Palis, now the President of the Brazilian Academy of Science, co-chairs the Indo–Brazil Science Council with me. Eduardo Krieger (a medical doctor and expert on blood pressure) was the President earlier. Besides being an honorary member of the Brazilian Academy, I am the fortunate recipient of two major Brazilian honours of which the Order of Scientific Merit (Grand Gross) is special.

An important benefit of being a visiting professor to a major university or attending an important international conference is that one gets to meet leaders in the profession. By just talking to some of them for a few minutes, one learns more about the latest advances than by reading the literature for days. These contacts are necessary to carry out frontline research. It is impossible to forget that in my own life some of the most important scientists I have met are at conferences. I remember meeting the famous theoretical chemist (Late) Charles Coulson at an international conference in Washington D.C in 1963. We sat next to each other in a bus for about two hours. My association with Nevill Mott started when I was a visiting professor at Oxford. When I was in Cambridge, I met Michael Klein (an outstanding computational chemist) in 1983 and he has since become a family friend.

There have been several occasions when I have felt woefully inadequate, with doubts about my ability and accomplishments. I get into this state of mind specially in international conferences, even after successfully delivering a plenary lecture. Often we see many friends and professional acquaintances at conferences, and witness heated discussions and casual comments on various topics. People are seen to be talking with great authority and self-confidence, although they may not be all that accomplished. Good scientists are seen bragging about their work or presenting it in a fashion that suggests that they are looking for a prize or some other major recognition. Coming from India, I have found it difficult to participate in conversation amounting to gossip about professionals in my area or about matters related to patents and activities in various laboratories. At such times, I have felt isolated. Why Indian scientists feel this isolation is perhaps because the main stream of science does not pass through here. (China is trying hard to attract this main stream.) Over the years, however, I have learnt to ignore such feelings of desolation and maintain equanimity.

While friendships and collaboration are important, what really increases visibility in the community of scientists are research publications, reviews and books. I have been lucky in this regard. It looks like many people read my research papers and books. They have been cited several thousand times in the literature, resulting in a large circle of professional friends and acquaintances. I remember a funny incident that happened recently. I was in a European airport after giving a lecture at an international conference. A gentleman came towards me and said "Hello, I heard your plenary lecture. I have read papers of another person with your name published in the 1960s. He must be dead by now". I had to tell him that I was the same person. The long innings and the large number of publications have indeed provided me a large circle of friends and colleagues all over the world. It is a wonderful feeling of belonging that keeps me happy.

While the various international connections and activities have had beneficial effects on my professional life, my association with the Third World Academy of Sciences, now called the Academy of Sciences for the Developing World (TWAS), has had the most influence on me in developing a proper attitude and having the right priorities. This Academy with its head office in Trieste, was founded in 1983 with Prof. Abdus Salam (Nobel Laureate in Physics from Pakistan) as the Founding President. I am a founding member of this Academy. M.G.K. Menon and M.S. Swaminathan are two other surviving Indian founding members. I have taken great interest in its activities since 1986 and have served the Academy in various capacities. In this process, I have made a large number of friends from the developing countries of Asia, Latin America and Africa. This would not have been possible through any other organization. More than meeting people, this Academy has revealed to me the real-life situation in the world of science. The situation in the LDCs or the least developed countries (there are over 40 of them) is so horrendous that it is difficult to imagine how anyone can do meaningful research or teaching.

I have visited some of the LDCs in Africa and elsewhere in my capacity as the President of TWAS in 2000–2007. I find that even in the best universities of the LDCs, the facilities are not much better than those in good high schools in the advanced countries. This saddens me. The conditions under which people in developing countries function even today, is depressing. This reminds me of my own situation 50 years ago, when I had no equipment and chemicals, and no money to carry out worthwhile research or to attend international conferences. The conditions in India that I faced in the 1950s were probably superior to those prevailing today in many of the developing countries. Unless we do something to help developing countries, particularly the least developed ones, to come up in science, I do not see how the future of the world will be a reasonable one. For the sake of peace, it is necessary to have equity in knowledge, if not equity in the economic sphere. It is for this reason that TWAS spends most of its efforts in improving capacity building in developing countries.

TWAS has a membership of around 800 from 90 or more countries. Around 12% of the members are from advanced countries and are elected to the Academy because of their involvement with developing countries. Election of a person as a Fellow of TWAS has become a major recognition in the developing countries. TWAS has held general conferences in various parts of the developing world, Abuja, Alexandria, Beijing, Dakar, Delhi, Durban, Kuwait, Mexico city and Rio de Janeiro being some of the cities of recent meetings. The most important programmes of TWAS are: giving awards to bright young scientists, providing fellowships for South-South cooperation and South-North Cooperation, grants for spare parts, supporting international conferences and so on. TWAS prizes are considered to be major international recognitions for scientists in the developing world. It is good to see that in the last few years, TWAS is recognized as the forum of the scientific community in the developing world. I feel extremely proud to be

Rao in Egypt with the Pyramids in the background

associated with it. It is indeed noteworthy that India is one of the major players in TWAS, with a large contingent of fellows and prize winners.

China, Egypt and India can boast of past civilizations which were magnificent. In Egypt, one goes to the tombs in Luxor and the Pyramids in Cairo just as one goes to the forbidden city and the great wall in China or the Taj Mahal and Hampi in India. A modern temple that has been created in Alexandria is the Alexandria library (Bibliotheca Alexandrina). Ismail Seregaldin has created this remarkable institution. It has something to show of the past and a direction to the future. I had the pleasure of conducting a workshop on nanoscience and technology there. We need an equivalent organization in India which will provide a proper exposition of the Indian heritage both ancient and modern.

There is still discrimination in the world. I experienced discrimination in my own life many years ago. When I went to the US in 1954, I was not served coffee in some of the coffee shops. I was not given a room in hotels in some cities. Derek Davenport of Purdue University often reminds me of our experience 55 years ago in Harry's bar next to the university campus. They would not serve us because I was there. Harry Belafonte, the great entertainer, performed at the university when I was a student. When Belafonte tried to go to the bar after the performance, he was not allowed entry. This may not happen today, but there are still problems affecting many of us. However good the work may be, a young inexperienced referee from an advanced country tends to assume that the work coming out of a developing country cannot be that good. Scientists from India still seem to face this problem, let alone scientists from countries in Africa. After more than five decades of publishing, I occasionally get referee reports where I am taught elementary science, with questions raised about simple procedures. The only thing that I am not asked is whether I brushed my teeth in the morning.

We must all help and promote individual research of quality in the poorest of countries because science is required in all the countries of the world, however small and however poor. It is for this reason that

TWAS has initiated a new research grant system for good scientists in the least developed countries. Sweden has supported this research grant programme in a big way. The C.N.R. Rao Education Foundation has established a prize for a leading research scientist from an LDC under the auspices of TWAS.

It was a pleasure to work with Abdus Salam when he was alive. We conversed generally in Hindustani. I have enjoyed working with Mohammed Hassan (from Sudan) who has been the Executive Director of the Academy for many years. Mohamed is like a brother to me, and has been a solid member of the TWAS family. After my tenure as President of TWAS, Jacob Palis has taken over as President and I continue to work on various aspects of the Academy as the immediate past-president.

As a representative of TWAS, I have participated in the Inter-Academy Panel and Inter-Academy Council which have all the science academies of the world as their members. Many reports and declarations have emerged from these bodies. Unfortunately, little of consequence is witnessed in the developing world because of reports of international bodies. In this respect, most organizations including the world conferences organized by UNESCO and other world bodies seem to have had limited success. Clearly, the developing countries will have to work hard for their future. It is encouraging that Brazil, India and China have emerged in the last few years as powerful nations. There are other rising nations such as Chile and Mexico. My only hope is that in the next two decades, a majority of the developing countries, specially the LDCs, will emerge from the present state to a new era where the power of knowledge can level the playing field with the more privileged. It will be wonderful if the developing countries can contribute 30% to the world scientific literature.

7 Reflections

I became 75 years old on June 30, 2009, 58 years after I first entered a research laboratory and 50 years after I embarked on an academic career. In 2011, I will celebrate the diamond jubilee of my research career. I have had a wonderful life doing science, and have enjoyed doing research even more after I turned 60. I feel that I have done my best work since then. In the last forty years or so, I have been elected to almost all the major science academies of the world and have been recognized in various ways by professional societies and others. As I am getting older, I am getting an increasing number of invitations to lecture all over the world and to be a visiting or adjunct professor in foreign universities. I have been a distinguished visiting professor at UC Santa Barbara and Berkeley for sometime, but I recently decided not to move out of my base in Bangalore for long periods. A university in the Middle East offered me an unbelievable salary if I could go over there for a year or two (or even for a month each year). I took no time to refuse the offer, though politely. Money has not been the attraction in my life (fortunately). I have received 48 honorary doctorate degrees from both Indian and foreign universities. I have found real solace and satisfaction in the research papers coming out of my laboratory. I have had dedicated research students at the Indian Institute of Science, Jawaharlal Nehru Centre for Advanced Scientific Research and the Indian Institute of Technology Kanpur. Most of my research has been done through students who have given their best during their stay in my laboratory. I have always had 12 to 15 Ph.D students working with me. Including Masters' students, project assistants and post-doctoral fellows, the size of my research group is generally 20 to 25. Around 140 young people have received Ph.D degrees while working with me. I have also guided the Ph.D research of several students even though I was not formally their research supervisor. I have thoroughly enjoyed working with students and other coworkers. I have been writing research papers all my life (almost every day of my life), and without even knowing, I have become an author of over 1400 research papers and 42 books. The publishing virus that attacked me in my youth seems to have had a powerful effect. The virus is harmless and even seems beneficial. I am told that our papers have been cited widely (close to 40,000 times), giving me a high H-index. What is more noteworthy, however, is the way my research subject has grown over the past few decades.

As I mentioned earlier, my students have been wonderful. These young people have kept me young. Working on a new subject makes me even younger in spirit. I would have liked to thank each of my students and other coworkers individually, but that would be impractical*. Some of my students have become well known scientists in their own right and are spread all over the world. Some of them are professors in Europe and USA and many of them are professors in India. A few of them have become leaders in academia. Amongst those who have worked in solid state and materials chemistry, I must mention K.J. Rao (inorganic glasses), G.V. Subbarao (metal oxides), S. Ramasesha (theoretical chemistry), A.K. Shukla (batteries and fuel cells), K.C. Patil (combustion synthesis), S. Yashonath (computer simulation), G. Sankar (catalysis and EXAFS), D.D. Sarma (electronic structure), T. Pradeep (nanomaterials), S. Vasudevan (Catalysis), A.K. Ganguli (superconductivity) and Ram

* A list of my scientific collaborators is given as an Appendix.

Seshadri (Chemistry of Materials). Surjit Singh and others worked on spectroscopy of molecules. Some of my students have come from very poor families and it has been a pleasure to see them advance and prosper. I have many academic grandchildren and great grandchildren. A few great great grandchildren have also started to show up. I have not had close research collaboration on a continuing basis with my students. The only exception is A. Govindaraj who joined me for his Ph.D work in 1991 and continues to work with me on nanomaterials.

I could have taken many patents, but I have not. I will rectify this in my next birth. Many countries seem to have exploited my research results to make industrial products as per a study conducted by the CSIR.

It often becomes depressing to deal with journals and referees. Some journals have become so holy that they refuse to consider papers for publication just based on physical inspection. Some referees are insulting and ignore the fact that the investigators have some abilities and experience. Referee-fatigue and journal practices discourage the young, in some cases, sufficient to render them inactive and cynical. After dealing with a few thousand referees, I am not giving up. I feel that it is best to publish in good professional journals without worrying about the artificial hierarchy of journals. As Professor Mott used to tell me, "if the work is good, people will read it". This is the best attitude indeed. I notice youngsters spending enormous effort to publish a paper in a specific journal of repute and then rest for a year or two out of exhaustion.

When I started working in solid state and materials chemistry in the early 1960s, I was often a subject of ridicule. My wife often reminds me of one or two senior professors in the country telling her how I was working in areas of chemistry that were not in vogue. "Tell him that he should change if he wants to succeed in India", they had advised. Fortunately, I had the strength to ignore the advice. Today, solid state and materials chemistry has become a major direction in main-stream chemistry and is pursued widely all over the world*. I am reminded of a famous poem by an American poet "Tall oaks from little acorn". When one sows a seed, a sapling comes out first. The sapling becomes a plant and the plant then becomes a tree. The trees give rise to a forest and the forest becomes a continent. There is no greater pleasure than seeing one's own subject of interest become a continent.

Not following a beaten track and going after a lesser known subject has yielded dividends. A poem by Robert Frost seems appropriate in this connection:

> Two roads diverged in a yellow wood
> And sorry I could not travel both
> I took the one less travelled by
> And that has made all the difference.

I am proud to be a chemist although chemistry is not the most fashionable or sought-after subject. Unfortunately, the chemistry we teach and the chemistry we do in the research laboratory are different. I started as a physical chemist which itself got recognized as a major branch of chemistry only in the early 20th century, one of the earliest physical chemists being Ostwald. I owe my inspiration to the more modern physical chemists, G.N. Lewis and Linus Pauling. The subject of solid state and materials chemistry, however, is not a derivative of physical chemistry, but makes use of a variety of facets of chemistry. My approach to the subject was initially based on physical chemistry (relating structure and properties, and studying phenomena in solid materials). I found that this was not enough if one really had to contribute to this field meaningfully. I had to add on many other aspects of chemistry to my efforts, and this has made me a better chemist. Divisions within chemistry and in science get blurred when working on advanced materials. In recent years, people have considered me to be an

* See the essay entitled "Chemistry of Materials: A Letter to a Young Friend" included in the Appendix.

inorganic chemist, a materials scientist and even a physicist. I must point out that the chemistry that I knew when I was young has changed a lot, just as the rest of science has changed in the last five decades. It is likely that someone who has not worked in the laboratory for a few years will find it strange when he reads present-day literature. Working continuously in solid state and materials chemistry for a few decades has been a great boon to me. I was pleased to see the tribute paid by Robert Kahn in his book, *The Coming of Materials Science*, where he states, "Rao had been active in several of the major aspects which, together, were beginning to define materials chemistry".

My family has been supportive in all my efforts. My parents gave me the freedom to choose the line I wanted. My wife has been a terrific partner and has taken care of all domestic matters. I have not gone shopping with her, but we have common interests in literature, music and theatre. H.C. Brown used to say that he did only two things in life: one was to do good research and the other was to keep his wife happy. I have strictly followed this principle. I have worked seven days a week all through the years, eating lunch at home only on Sundays. Happy family life is an important criterion for success in professional life (of scientists, at least). Our home has always been blessed with wonderful people who help us in the kitchen and in other household chores. In Kanpur, we

With Indu at JNCASR

With Prof. and Mrs. H.C. Brown

had Kamla and her family. In Bangalore, Frances has been with us for nearly three decades and has been a great support to us all.

Our daughter Suchitra is nicely settled in her Pune home and is working as a teacher. Our son-in-law Dr. K.N. Ganesh is doing well as a scientist and heads the IISER in Pune. Sanjay works in education technology. Our grandchildren are growing up nicely. Our grandson, Kartik, is interested in literature, theatre and music. We discuss books

With Sanjay and Kartik

Suchitra and Ganesh with Suguna and Kartik

With Indu and Sanjay

(specially fiction) regularly. Our granddaughter Suguna is going to school and is showing signs of becoming an artist. The main holidays that I have had are during conferences, but we go to the Himalayas or to the jungles of India regularly. I have enjoyed these trips with my children and grandchildren. We watch movies (DVD's) at home on some Sundays with Sanjay and Kartik. We depend on Kartik for the choice of movies. I read general literature* as much as I can, and cook occasionally for my family. Cooking is a wonderful hobby because when one cooks, one forgets everything else in the world. Unlike research, one can see the results of the efforts right away.

A trip with my family during the Christmas vacation of 2004 to Sri Lanka can never be forgotten. We were in a nice hotel, south of Colombo, when Tsunami hit the island on the morning of December 26. We were on a boat on the river when this happened. We could feel the terrific surge of water. The boatman became panicky, left us somewhere and ran away. We could see crocodiles and turtles being washed on to the shores. We had to return to the hotel, which we managed with great difficulty, with me carrying my little granddaughter on my back. She kept saying that she was not afraid since her grandfather was there.

* Historical novels and scientific biographies are my preferences. P.D. James is my favourite author for mysteries.

Everywhere there was water, and the roads were full of boats, debris, all kinds of furniture and equipment from the hotels. The scenario in the hotel was shocking. We waded through water and reached the foyer. The waves from the sea had sucked away the furniture, people and other contents from the rooms on the ground floor. Some of the people that we had seen at breakfast were no more. People who went for walks along the beach were not to be found. Our room was on the second floor and our belongings were in tact. When all this was happening, there was no way to contact the outside world. We were marooned and I was praying. I was really touched when I received a call from Tony Cheetham all the way from California. Tony had seen the disaster caused by the Tsunami on CNN. That was the only call that I received on my mobile. We soon found out that the train that was to arrive from Colombo had been washed away and was somewhere in the ocean. As we were all sitting in the corridors of the hotel not knowing what to do, Martin Jansen (our dear friend from Germany) came looking for us. Martin and his wife Monica were also enjoying a vacation there and Martin had gone snorkeling in the ocean that morning. It is unbelievable that he swam for a mile to the shore after the Tsunami hit us. Eventually, the hotel (which wanted all of its occupants to vacate) found a bus to take us to Colombo by a circuitous route. The bus was full of seminude foreign tourists who had lost their baggage and friends. We arrived in Colombo early next morning. I found an autorickshaw which agreed to take us to Colombo airport. Thanks to the crew of Jet Airways, we were accommodated in their rooms in a hotel near the airport. We reached Bangalore later by a Jet flight. Having witnessed the destruction caused by the Tsunami, I wonder whether we can ever be prepared sufficiently to face disasters which occur with such speed.

Returning to ordinary life, I must confess that because of the schedule I keep, I have had no time to pull down others or think ill of others. My wife has often told me that she has never caught me lying any time.

I have not found the need to lie, although speaking truth may not always be pleasant. I also bear no ill-will towards anyone and have not had the time to dislike anyone intensely. I have been reprimanded by people for rushing to the laboratory directly from the airport. I stopped doing so a while ago. I, however, still feel guilty on days when I am unable to put in a few hours for research. Having a packed professional life has had some disadvantages. I had no time to visit my relatives and friends. I have a large number of relatives right in the city of Bangalore, but I do not know where they live. Many of them are first cousins. I have not attended birthdays and other celebrations and feel terribly guilty about my behaviour. Indu jokes that I have attended only two marriages, my own and my daughter's. She takes care of all the social obligations, answering embarrassing questions regarding my absence. I wish that I did not have to do this. Unfortunately, when someone has a serious purpose in life, and wants to accomplish something worthwhile in India, it becomes difficult to do anything else. In experimental science, it is far more difficult to accomplish in India than in advanced countries. Even today, I am short of some of the essential facilities. For example, we do not have synchrotron (X-ray) or neutron diffraction facilities in the country. We occasionally get slots to use facilities abroad. I have had to worry about every aspect of my laboratory for the last 50 years. When I started my research, there was no good electric supply in the city. I really could not carry out experiments which required a constant supply of electricity. This problem continues to plague me. Even now, I have to worry about electricity and water in the laboratory. Keeping sophisticated instruments in good condition is another challenging task. A scientist does not have the luxury of being relaxed, enjoying life the way most people do, if he/she has to really accomplish something worthwhile. I hope that people will understand why I want to be left alone at home when I am not in the laboratory. I sincerely apologize to all my relatives and friends for having been a constant absentee.

What has kept me company throughout has been classical music which I listen to every morning and evening. Music has become part of my life. In the last few years, I start the day with music early in the morning. This is possible since I wake up very early in the morning for a walk. Some of my favourite Hindustani vocalists are Bhimsen, Rashid Khan, Jasraj and Ajoy Chakravorty. MS, Madurai Mani Iyer and Maharajapuram Santanam used to be my favourite Karnatak music vocalists. Among the instrumentalists, I like Bismillah, Vilayat, Hariprasad and Shivkumar. I also enjoy western classical music, opera and Jazz.

Simba, our golden labrador, sits with me and my wife to listen to music in the evenings. Bilbo, our doberman, also used to enjoy music. Our dogs have been great company and are always with us for morning coffee and again for evening coffee. They sit with us upto the 9 p.m. news in the night. Bilbo behaved like a gentleman and was always there to greet me when I got home in the evening, with a big grin (believe it or not!). I had to just pat him on his head. The day before he died in a hospital, he began barking when he heard my voice. It seemed he just wanted to be patted by me before he went to sleep.

With Suguna and Simba

Rao giving a lecture to children

Although research has occupied most of my time and effort for the last five decades, I have enjoyed teaching. Some years ago, I wrote a textbook on chemistry and taught in the classroom. In the last few years, I have attempted to reach young students all over India through direct contact programmes. I give a lecture with power-point presentation followed by a multimedia presentation by Indu. Indu has helped immensely in this effort and we have travelled all over India for this purpose. In Karnataka, these programmes are conducted in Kannada and our roadshow has been popular and sought after. I seriously believe that direct contact with young children is the only way to excite them about science. Three years ago, I lectured to a few hundred talented young students at the Asia Science Camp in Bali. Several Nobel Laureates were lecturers as well. It is wonderful that we have started organizing similar camps (Vijyan Jyothi Shibir) in India. Our educational effort has resulted in books and CDs, the most important of them being *Learning Science* (a four-part book) authored by Indu and myself. *Learning Science* is being published in various languages by the National Book Trust (NBT). It is meant for young students to learn on their own. It covers most aspects of elementary physics, chemistry, biology and earth sciences, in addition to many stories. Another book entitled *Understanding Chemistry* has come out in Indian and international editions. NBT will soon bring it out in several Indian languages. I have just written a booklet called "Nanoworld" for use of children and the general public.

Indu and I have enjoyed our educational efforts. When I received two major awards few years ago, Indu suggested that we donate a good part of it to set up an education foundation to promote science education in various ways. The foundation awards good science teachers in the country and the best students in IISERs. It also supports a bright student from IIT Kanpur, besides awarding prizes to the best researchers in BHU and elsewhere. There is an award for an outstanding scientist from a least developed country administered by TWAS.

Working through the years, I have often felt professional loneliness. This feeling is easy to pick up when working in India, a situation that was more true in the early years of my research. To strive on, despite the various problems, personal, professional and environmental, one needs strength and determination. I will be dishonest if I did not recognize the immense support of my wife and family, and the great faith that I have had in the almighty as the real source of the courage and energy needed for my efforts*. Otherwise, I do not see how a boy from Basavanagudi could have come this far.

I have worked with a large number of colleagues and coworkers in India and many have been good friends. They come from all parts of the country and speak different languages. As one progresses in science, one also tends to lose some friends. A peculiar kind of gap develops if he/she becomes single-minded about research. I will, however, mention a few old friends for their special qualities, but this is not to imply that those not mentioned here are less dear to me. I am doing this at the cost of being misunderstood by some of my friends, specially of the younger generation. A few of the friends whom I have known for decades are Roddam Narasimha (a schoolmate), Arcot Ramachandran (Purdue

* See the essay on *Science and Spirituality: Personal Reflections* included in the Appendix.

alumnus and well-wisher), S.K. Joshi, P. Rama Rao, T.R. Kasturi and R. Kumar (last two from old IISc days). M.V. George from Kanpur days has been a loyal friend for over four decades. He is a gentleman and a god-fearing person. I work with George in organizing workshops for science teachers and post-graduate students in Kerala. (I love Kerala for its cuisine and its wonderful environs, specially the back waters. It is truly god's own country). M.M. Sharma (whom I first got to know in 1967) is someone I admire for his high integrity and honesty. He is the best chemical technologist that I have known. He is a friend with impeccable qualities. It has been a pleasure to see Ramesh Mashelkar (whom I first met when he joined NCL as a young chemical engineer) rise in stature and influence. I have wonderful friends all over the world and I should specially mention Tony Cheetham (now in Cambridge) and Mike Klein (in Philadelphia). Martha Greenblatt (Rutgers), Reshef Tenne and Joshua Jortner (Israel), Martin Jansen, Herbert Roesky and Achim Mueller (all from Germany), Dick Zare (Stanford), Ahmed Zewail (Caltech) and others have been some of my dear friends, besides those whom I mentioned earlier in the book.

I first met M.G.K. Menon in the late 1960s, but got to know him closely in the late 1970s. He is one of the most brilliant persons that I have known. He would have scaled great heights in physics if he had decided to continue in research, but he sacrificed everything to serve the country in various ways. He first became secretary of electronics after moving from the directorship of TIFR and then served the government of India in several capacities. I remember that he joined TIFR in Mumbai after rejecting the offer of a good faculty position in Columbia University. Arcot Ramachandran left IIT Madras to become secretary of the DST. He was an outstanding secretary, and later served the UN in the Habitat programme as Under-Secretary General. He was a forthright, no-nonsense administrator and has been a loyal friend. S. Varadarajan left the chairmanship of major public sector undertakings to become secretary to the government. He has served many organizations in the country. He is one of the most generous persons I have known. I sometimes worry that these gentlemen, and probably some others too, have not been properly rewarded or compensated by the Government. I mentioned earlier about Prof. T.R. Seshadri who did not have enough to live on comfortably after his retirement. It is about time that the country ensures suitable recognition and protection for scientists.* G.N. Ramachandran had to wait so long to be recognized by the Royal Society, London. Did India recognize him sufficiently? When I met Ramalingaswami (the most outstanding medical scientist we have ever had) a few days before he died, he wept and asked me, "Ram, why did the country not find me fit for due recognition?". (An annual lecture has been established in his name at JNCASR).

Raja Ramanna was a dear friend for many years. He enjoyed good things in life and had a fine sense of humour. After his tenure in atomic energy, he lived in Bangalore and was associated with NIAS. He could have lived for some more years, but he ignored his health. Just a few days before he passed away, he asked me to meet him for lunch at the Taj West End (which we did occasionally) and we had an enjoyable afternoon. A few months earlier, when we were in Ahmedabad, he had difficulty walking to the meeting room. He took rest that day and a few weeks later he was wheeled into an operation theatre in Bombay to get a pacemaker fixed. When the doctor assigned to the procedure went out to take a phone call, Ramanna decided to walk out of the hospital since he somehow understood from the youngsters in the operation theatre that the procedure may not be all that useful. Ramanna and I used to meet frequently to talk about various matters, but he is no more.

I adore my heroes in science. By and large, my heroes have been those who have worked throughout their life. My heroes from the west have been Michael Faraday, G.N. Lewis, Linus Pauling and Nevill Mott.

* We rarely talk much about the work of living scientists and always recall the work of those who are no more.

From India, they are J.C. Bose and C.V. Raman. I would like to say a few words about my heroes.

My hero from the 19th century is Michael Faraday. It is inspiring to ponder over his life. How did a person with three-years of schooling publish over 400 research papers single handedly, and make so many original discoveries which would have probably earned him five Nobel prizes if he were in the 20th century! It is mind-boggling. Yet, he took pains to talk to children and the public. He refused the presidentship of the Royal Society since it would take him away from the laboratory. He worked till the day his health failed. His working habits were extraordinary. He recorded everything that he did in the laboratory meticulously in note books. It is interesting that on December 24, 1833, he carried out experiments in one area. On December 26, he researched on another topic. There is no record of his work on December 25, since it was Christmas day.

J.C. Bose carried out his seminal work on microwaves in the 19th century. After his return from Britain with degrees in physics, he started teaching in Presidency College, Calcutta. After a few years, he decided to embark on research. Within a year, he discovered the coherer and publicly demonstrated how microwaves go through space and matter. (Amazing! In one year.) This work, published in 1895, marks the beginning of telegraphy. Yet, similar work carried out two years later by Marconi was given the Nobel Prize. J.C. Bose continued his research work on plant physiology and seemed to take ups and downs in life philosophically. Some of his activities in research are puzzling. For example, he used lead sulfide crystals in his study of microwaves. On what basis did he pick lead sulfide? Solid state physics had not developed at that time. It is amazing that J.C. Bose survived the highly oppressive administrative attitudes of the British those days. When Lord Rayleigh (who had given great support to J.C. Bose) visited Calcutta, he wanted to see Bose's laboratory. Bose took him around the laboratory, only to receive a letter from the principal the next day, admonishing him for taking Lord Rayleigh around in the college without his permission. When J.C. Bose wanted a small research grant, the initial reaction of the local administrator was one of horror. How can a native scientist ask money for research, instead of using savings from his salary? To me, J.C. Bose was the first modern scientist of India who should have been India's first Nobel Laureate. J.C. Bose probably survived those days because of his spiritual outlook. He became an F.R.S late in his life at the age of 62. He was a close friend of Rabindranath Tagore and this relationship perhaps played an important role in his attitude to life. J.C. Bose gave away all his life's savings to establish the Bose Institute in Calcutta.

C.V. Raman worked in Calcutta when J.C. Bose was still alive. Raman started his research career in Nagpur working in the local science college during evenings and spending the day in his office as an accounts officer of the British government. His early work was on Indian musical instruments. In Calcutta, he worked at the Indian Association for the Cultivation of Science (IACS) and discovered Raman effect (Nobel Prize, 1930). His genius was recognized by Asutosh Mukherjee who invited him to be Palit professor of physics in the University of Calcutta. Raman trained a number of students in Calcutta and shifted to IISc, Bangalore, in 1933 as the first professor of physics. Raman was truly a great man who should be viewed in totality. One should not exaggerate his idiosyncrasies. If he liked someone he would go to any lengths for the person. He did not like phonies. There is the story of a student who had picked a grating by his hand leaving his fingerprints on it. Raman made him pack up his stuff in the student hostel, and personally took the student in his car to the railway station. He left him there and asked him not to come back. What is wrong with that? Just look at the galaxy of scientists Raman produced. GNR was probably his best student.

J.C. Bose found support from Lord Rayleigh, and C.V. Raman had the encouragement of Lord Rutherford. Rutherford ruled the world of physics at that time and the Cavendish laboratory in Cambridge that he headed was the Mecca of physics. Rutherford was a student of J.J. Thompson (who had discovered the electron) and succeeded him as the Cavendish Professor. Rutherford is known for his work on the structure of atoms, but his bigger contribution was in nurturing a large number of young people to carry out research at Cavendish on important problems. Some of them received the Nobel Prize. Rutherford took pride in the accomplishments of his coworkers and students, most of whom became famous. Cockroft, Walton, Chadwick, Blackett and Kapitza are some of Rutherford's famous coworkers. Bohr too considered Rutherford as his teacher. There was a joke about Rutherford's Laboratory at that time. It seems that taxi drivers were hesitant to take people to the Cavendish Laboratory. They would say, "If we go there, Rutherford will put us to work and we will end up with a Nobel Prize". When someone complimented Rutherford by saying that he was riding the waves, Rutherford apparently quipped, "Who makes the waves?".

Nevill Mott joined the Cavendish Laboratory in Cambridge as a research student, but to do theory. This meant working a little far away from the Cavendish laboratory. (Theoreticians were kept away in Silver Street by Rutherford.) Mott became a professor of theoretical physics at a very young age in Bristol. He built a formidable group in condensed matter physics there before returning to Cambridge after a few years as the Cavendish Professor. Besides his accomplishment as a physicist, Mott possessed fine human qualities. He brought many refugees from Nazi Germany to UK and helped them in various ways. He was the first one to offer Hans Bethe a position in Britain when he came from Germany as a refugee. Mott was a simple man. He would stand in the long line for tea and lunch. I once visited him when he was nearly 90 years old. I saw him bent down at the table. When I asked him what he was doing, he showed me the proofs of a manuscript that he was correcting. I cannot forget the way he came to the Cambridge bus station to pick me up in 1974. Many people have told me of similar experiences. An unusual trait of Mott relates to the way he conversed on scientific matters. He would suddenly utter a word like beryllium or mercury when listening to a presentation. It would take many days for one to realize what he meant. He had an uncanny ability to relate one topic or observation to another. Mott was never hesitant to ask simple questions. He was asking questions till the last day.

I saw Linus Pauling for the last time at a meeting of the US National Academy of Sciences when he was felicitated on turning 90. After the citation was read out, Pauling walked up to the stage like a soldier and gave a short speech. I talked to him for a few minutes during the coffee break. He was fully active and agile. Pauling changed chemistry and chemical education. During his career, he taught freshman chemistry and authored path-breaking books on general chemistry. He also wrote the first book on quantum mechanics. He received two Nobel prizes and would have probably received another one for the DNA structure if only he had seen the x-ray photographs of Rosalind Franklin in London. Unfortunately, he was not allowed to go out of the US at that time. What is amazing is that he continued to be brimming with new ideas even after his 80th year. The question he raised on quasi crystals towards the end of his career triggered much research on the topic. It is not necessary to be always exact or correct. What is more important is that one comes out with new ideas. Pauling generated many ideas on chemical bonding, molecular structure, crystal structure, protein structure, sickle-cell anemia and so on. Once he was asked how he got a great idea. He just said, to have a great idea, one must have several ideas to start with. Pauling was fearless and fought for the ideals and principles that he believed in.

I would like to end this narration with a few personal observations on certain aspects related to science in India. India has seen unprecedented development in the last few years resulting in a significant increase in the GDP. There is a large middle class population in the country and the upward mobility of people in their economic status has been noteworthy. There is also much more money available for science and technology as well as higher education. In cities like Bangalore, industries and population are growing at a fast pace, but the infrastructure is unable to cope with the growth. The garden city of Bangalore is gradually becoming a garbage city. Too much importance to areas such as information technology, banking and management has resulted in an imbalance in the education sector, with a majority of the students taking up banking, business and such subjects. Science has suffered badly. Family and societal pressures are driving young people to professions, not necessarily of their choice. Over emphasis on sensex, foreign investment, GDP and such matters in the media and elsewhere has affected the attitudes and value system in our society. This is certainly not good for the country. We cannot have cities without souls. If India has to be a major player in the world, it would require top-class manpower in all sectors, and progress in various fronts. The science sector is important not only for its own sake but also for the development of innovative technology and industrial development. I do hope that good sense will eventually prevail in our society.

Big science has received undue importance in the country. For example, atomic energy and space get mentioned all the time. In reality, the world is being pushed forward more by small science. Accomplishments in small science benefit mankind much more. Unfortunately, those of us who do small science in a big way have to play the second fiddle. Some of the senior scientists from big agencies do not really understand or appreciate the difficulties of scientists in educational institutions. They tend to show benign indifference and their ideas of what constitutes excellence deviate widely from world view. I must squarely blame politicians, planners and administrators for giving excessive importance to big agencies and not providing proper moral support to scientists who work in small laboratories and educational institutions. It is not enough to throw money at good scientists. It is also important that due appreciation is given for their contributions. They should be treated with empathy and understanding.

A serious problem many of us face relates to the conflict between doing administration and science. This conflict has resulted in a number of good scientists being lost to science by becoming administrators at an young age. It is extremely difficult to be involved in administration and science at the same time. I admit that my second term as the director of the Indian Institute of Science resulted in diminishing the quality of my research. If I had not taken up the second term, I would have probably been way ahead in my research. It is my view that those who really wish to do outstanding research and gain credit through science should avoid administration. If they need to take up administration, then they should restrict it to something small, related to their expertise and interest. Note that administration takes up all available time. It is difficult to say no when one becomes an administrator. I am most grateful that I had the opportunity to recover from administration after giving up the directorship of IISc at the age of 60.

Many of us, specially the younger people, travel too much. In the early days when I started my career, there was no money to travel and it was impossible to go abroad. Now that it is easy to go abroad, many are abroad most of the time. If not abroad, they will be somewhere else in the country. There is a joke in IISc and in the JNCASR, that we have many distinguished visiting professors! Many years ago, I decided not to accept memberships of selection committees, examinerships, and so on to minimize travel within the country. Nowadays, I do not deliver convocation addresses, inaugurate meetings or accept many lectureships. I go by the principle that I should be in my laboratory for at least 20–22 days in a month on an average. This has helped me professionally. At least in experimental science, it is not easy to keep travelling while also hoping to accomplish great things in the laboratory.

Everyone knows that it pays to do hard work and yet very few work hard. In the Indian context, I feel that most of us do not work hard enough. The great scientists that I have known worked very hard. Several factors come in the way of our work in the laboratories. In the institutions where I have worked, the average young person does not work seriously for more than four to five hours a day. In my own experience, the modest success that I may have had in research is mainly due to hard work.

A common feature amongst some of my brilliant colleagues has been to spend much too much time solving puzzles, and proving that they are clever. Some become good in dealing with stocks and shares. Unfortunately, science loses them. Today, competition is so severe that it is becoming almost impossible to get noticed. In science, competition is tough, since one does not see the other competitors (unlike in a running race). One often gallops to remain stationary.

While talking of habits and individual qualities related to science, I feel that envy and selfishness can get in the way of good research. These qualities become more magnified in underdeveloped countries. I often think of Rabindranath Tagore's poem in Gitanjali:

> I came out alone on my way to my tryst.
> But who is this that follows me in the silent dark?
> I move aside to avoid his presence, but I escape him not.
> ...He is my own little self, my lord, he knows no shame;
> But, I am ashamed to come to the door in his company.

I realize that it is impossible to lose all selfishness. We have to be selfish enough to get results, but we have to make accommodation for others to do as well or better. Generosity is essential for science to thrive on a large scale. It is also important to ignore trivialities. Many of my colleagues, particularly the young, get worried about every small thing. Unless one stops worrying about trivial matters such as the price of vegetables, some new circular or something on the bulletin board, it is difficult to commit oneself to any creative activity. As Tagore wrote in Gitanjali, "Give me the strength to raise my mind high above daily trifles". How true! Another quality that hinders creative work is a rebellious temperament. When a person becomes rebellious and starts squabbling with colleagues or with the establishment, his/her research performance gets seriously affected. Some brilliant people have lost out in science and in other creative endeavours because of their involvement in trivial matters or by being excessively critical.

Many great scientists have said that picking the right problem is a more difficult task than doing it. This is indeed so. I find some of my brilliant friends working on "standard" problems and in crowded areas. Such work may result in a few papers, but may not have much impact. There is also a tendency amongst scientists to be repetitive. Students in IIT Kanpur had many jokes about scientific research. One of them goes as follows. There was a scientist working on a bug with four legs. He had taught the bug to jump whenever he shouted "jump". He cut off one leg of the bug, but the bug jumped when he shouted jump! Then he cut off the second leg. The bug still jumped when he shouted jump. He continued this study until no leg remained. When he shouted jump, the bug did not jump. He came to the important conclusion that the bug does not or cannot jump when all its legs are cut off. There was another scientist in competition, working on the same problem. He also found that the bug did not jump when all the four legs were cut off. He, however, came to the conclusion that the bug became hard of hearing when all the legs were cut off.

There is a tendency in India (probably in other developing countries as well), for anyone to claim to be a scientist. There are administrators who sit in their offices, but would not like to be denied their position as scientists. Of course, I do not mind calling them scientists, but they should in return give due respect and consideration to active scientists. It is common to find people who are not properly qualified and have not contributed to science in any way, talking more about science than those who actually work in the laboratories. It may be best to ignore this trait as a folly of individuals. However, one cannot but be amused by the way many of our politicians and film actors insist on being addressed as "doctors". I remember, Mr. K.R. Narayanan (former President of India) writing to me, not to address him as Dr. Narayanan, since he had not earned it.

I notice an undesirable trend wherein institutions slowly decay in quality. It is necessary for us to fight such decay. Good traditions of our institutions have to be maintained for the sake of posterity. In this context, I feel disappointed that organizations such as the Indian Science Congress and even our science academies have failed to meaningfully contribute to the improvement of the quality of scientific research and education. We cannot afford to lose our good institutions at any cost and we should do everything in our power to maintain the high quality of these institutions, because they are our national assets. What would we do if we did not have places like TIFR and IISc?

Senior scientists have a special responsibility. We have to be kind to young people and support them as much as possible. My own experience in the early days has made me realize the importance of this relationship. I have tried my best to be useful and helpful to as many young scientists as possible, but it is humanly impossible to be of help to all. I have had some unusual experiences with my own students and other young colleagues. Some of them become arrogant and cynical. We do not hear them saying "thank you" to anyone. Dragging down others has become a pastime of some. I cannot forget that G.N. Ramachandran did not enter the building of the Molecular Biophysics Unit that he built at IISc. I am reminded of Mark Twain's famous comment. When Mark Twain was told of someone who was talking badly of him, he remarked "It's amazing. I never helped him". Yet, it is essential that senior scientists have an affectionate approach to young people, and be role models. Senior scientists should compete with the young by producing good science. They must play a role in ensuring that the younger generation does not lose national pride which is important for the future of the country.

The future of science in India depends crucially on the pride that we take in our country. China seems to have succeeded eminently to make the younger generation proud of their country and of their role as scientists. Unfortunately, there appears to be a change in the value system of our society wherein parents take pride in their children being far away (in contrast to the days when my parents were so happy that I was not far away).

What I have tried to convey in the few preceding paragraphs was motivated by my intense desire to see that India becomes a scientific power*. Although we are not doing as well as we should in terms of both quantity and quality of scientific research, I see no obvious hurdles to reach greater heights. All that is needed is dedication, hard work and a better choice of problems. With proper attitudes and aspiration, we should be able to contribute 10% or more to the global research publications in another few years. At least 5% of the top 1% of the research papers in the world should be from India (instead of the present 0.5%). We should produce at least 10,000 Ph.Ds per year. In subjects like computer science, we need more Ph.Ds than service providers. In addition to blue-sky research, we should work on important problems

Happily talking to school children

* See the essay *Science in the Future of India* included in the Appendix.

arising from the needs of the nation and demands of the industry, specially where we have a unique advantage. Innovation should get top priority in all our efforts. If we succeed, we can be one of the largest knowledge providers in the world. Being a young country, we could dominate science in the world. I will not be alive when such progress takes place, but I do sincerely hope that it happens.

I end by recalling how satisfying my life has been. I have had a wonderful profession matched by an equally good family life. I have enjoyed my life as a scientist. I cannot think of a better way to live and I am getting to be happier as I am getting older. I have no regrets and would gladly do it all over again. I must express how grateful I am for the kindness, generosity and grace that have been bestowed on me in this great country and elsewhere, and for the innumerable doors that were opened for me.

"Man's reach must exceed his grasp,
else what's a heaven for"

Epilogue

Doing Science as a Way of Life*

C. N. R. Rao

Scientific pursuit is essentially a creative urge. Science is also an important part of our culture. Science is a demanding master, not easy to satisfy. Only those who completely submit to it seem to reap the benefits in terms of intellectual excitement and satisfaction. In this context, it is beneficial to recall the lives and work of great ***men who accomplished*** *against* ***all odds*** *and would not give up easily. What is also wonderful is that there is no limit to scientific pursuit. It is thrilling to witness the way ideas which start off as a mere grain of pollen develop to become forests and continents. Nothing can be more rewarding to a scientist than achieving a state where doing science becomes a way of life.*

Bernal wrote many years ago that 'Man is occupied and has been persistently occupied since his separate evolution, with three kinds of struggle: first, with the massive unintelligent forces of nature, heat and cold, winds, rivers, matter and energy; secondly, with the things closer to him, animals and plants, his own body, its health and diseases; and lastly, his desires and fears, his imaginations and stupidities'. I cannot improve on Bernal's statement, but it is my feeling that whatever be his struggle, the thinking man has always exhibited the distinct characteristic of compulsive occupation with unfettered intellectual explorations to satisfy his philosophical urges. Science is one of those urges. In social science usage, the term 'Science' denotes the systematic, objective study of empirical phenomena and the resultant bodies of knowledge. According to this definition, many disciplines related to social sciences should also come under Science. Science as a human activity is itself a subject of social science investigation. One must admit therefore that as an intellectual effort, science cannot be regarded as something apart. It is an important part of our culture. Science permeates our thinking today and it would be unimaginative to completely separate it from what we normally refer to as humanities and social sciences. Natural science does not simply describe or

*** Reprinted from CURRENT SCIENCE, VOL. 64, NO. 5, 10 MARCH 1993, P. 288-293**

explain Nature. It is a part of an interplay between Nature and man. As Heisenberg stated 'Science describes Nature as exposed to our method of questioning'. One aspect that distinguishes science from other scholarly pursuits however is that much of science is based on experimentation.

It is not my purpose to discuss the nature of science or of scientific thought here. Neither am I going to dwell on science and humanism or spirituality. There is much written about such matters by wiser men. I shall concern myself with those aspects of science and scientific pursuit that are necessary for the purpose of this presentation, which deals with doing science as a way of life. In such an endeavour, I shall make use of the thoughts of great personalities who have contemplated about these matters, besides recording my personal reflections. Before I proceed with the theme of my lecture, I would like to briefly mention some of those aspects of science that have made great impression on me.

Some features of science

Imagination plays a vital part in science, but has certain constraints. As stated by Feynman, 'whatever we are allowed to imagine in science has to be consistent with everything else that we know ... We cannot allow ourselves to imagine things which are obviously in contradiction to the known laws of Nature. Our kind of imagination is quite a difficult game. One has to have the imagination to think of something that has never been seen before, never been heard before. At the same time, the thoughts are restricted in a straight jacket, so to speak, limited by the conditions that come from our knowledge of the way nature really is. The problem of creating something which is new, but which is consistent with everything which has been seen before, is one of extreme difficulty'. At the same time, the difficulty with science is often not with the new ideas, but in escaping the old ones. A certain amount of irreverence also seems to be essential for creative pursuit in science.

Science does not allow exceptions. Without this feature, there would be no determinism in science or there would be no science at all. Another interesting feature is that 'full explanations are often seized in their essence by the scientist long in advance of any possible proof'.

Unanimity of opinion is not necessary in science. While unanimity of opinion may fit a religious organization, a variety of opinion is absolutely essential in science. Science is the only area which has not produced sects (although there are disciplines and schools in science). It is founded on analysis and synthesis. It does not occupy itself with probable truth. More importantly, it has the same method in all parts of the world. In questions of science, as Galileo said long ago, 'the authority of a thousand is not worth the humble reasoning of a single individual'. 'Freedom is the first-born daughter of science' as Thomas Jefferson put it. Yet, there is no democracy in science. We cannot say that some second-rate person has as the same right to opinion as Einstein or Pauling. It is useful to remember the motto of the Royal Society, 'Nullius in Verba' (we do not take anybody's word for it).

Jacob Bronowski has written, 'By the worldly standards of public life, all scholars in their work are, of course, oddly virtuous. They do not make wild claims, they do not cheat, they do not try to persuade at any cost, they appeal neither to prejudice nor to authority, they are often frank about their ignorance, their disputes are fairly decorous, they do not confuse what is being argued with race, politics, sex or age, they listen patiently to the young and to the old who both know everything. These are the great virtues of scholarship, and they are particularly the virtues of science'.

Science has nothing to be ashamed of, not even in the ruins of Nagasaki'.

Science and aesthetics

That Science admits aesthetic criteria just like the arts, has been expressed by a number of great scientists and mathematicians. It has been said that science has something intrinsically poetic. Science has been considered to be a meeting place of two kinds of poetry; the poetry of action and the poetry of thought. According to Bohr, 'when it comes to atoms, language can be used only as in poetry. The poet, too, is not merely so concerned with describing facts as with creating images'.

When Einstein was reproached that his formula of gravitation was longer and more cumbersome than that of Newton, he apparently said that 'if one were to describe the truth, leave elegance to the tailor'. Yet, Einstein's equation has been considered by many to be complete and beautiful. According to Hardy, 'Beauty is the first test; there is no permanent place in the world for ugly mathematics'. The place of beauty in equations was expressed by Dirac many years ago. According to Dirac, 'it is more important to have beauty in one's equations than to have them fit experiment. If one is working from the point of view of getting beauty in one's equations, and if one has really a sound insight, one is on a sure line of progress. If there is not complete agreement between the results of one's work and experiment, one should not allow oneself to be too discouraged, because the discrepancy may well be due to minor features that are not properly taken into account and that will get cleared up with further developments of theory'. Dirac was a genius and I would not question his statements or convictions. I, however, find it difficult to accept what he said literally, although I appreciate the spirit of it.

The quest for beauty in science has found excellent expression in the writings of the astrophysicist, S. Chandrasekhar. While examining the question as to the extent to which the quest for beauty is an aim in the pursuit of science, Chandrasekhar quotes Poincare. 'The scientist does not study nature because it is useful to do so. He studies it because he takes pleasure in it; and he takes pleasure in it because it is beautiful. If nature were not beautiful, it would not be worth knowing and life would not be worth living ... I mean the intimate beauty which comes from the harmonious order of its parts and which a pure intelligence can grasp.... It is because simplicity and vastness are both beautiful that we seek by preference simple facts and vast facts; that we take delight, now in following the giant courses of the stars, now in scrutinizing with a microscope that prodigious smallness which is also a vastness, and, now in seeking in geological ages the traces of the past that attracts us because of its remoteness'. Commenting on Poincare's perception, Sullivan states that 'it is in the aesthetic value that the justification of the scientific theory is to be found and with it the justification of the scientific method .... The measure in which science falls short of art is the measure in which it is incomplete as science'.

Chandrasekhar adopts two criteria for beauty. The first is the criterion of Francis Bacon that 'there is no excellent beauty that hath not some strangeness in the proportion'. The second one, due to Heisenberg, states that 'Beauty is the proper conformity of the parts to one another and to the whole'. Clearly, as Chandrasekhar says, each of us, in our own modest ways, can achieve satisfaction in our quest for beauty in science.

I must express some misgivings about taking arguments based on beauty or such other aesthetic factors much too seriously since they may not be conducive to vigorous research demanded today. After all, we are no longer in the golden age when one was mainly concerned about universal physical laws for all time. Furthermore, to limit beauty to mathematical equations alone would be unfair to science. There is beauty in the architecture of molecules and materials as well as in the pathways of transformations. There is also beauty in properly designed and well-executed experiments. There is beauty even in the way science works and develops. 'It is the great beauty of our science that advancement in it, whether in a degree great or small, instead of exhausting the subject of research, opens the doors to further and more abundant knowledge, overflowing with beauty and utility', as stated by Faraday.

Science and utility

Science has its uses, but the basic urge to do science is not for utilitarian purposes. It was the view of Helmholtz that 'whosoever, in the pursuit of science seeks after immediate practical utility, be rest assured that he seeks in vain'. There is some truth in it because while pursuing science it would be difficult to fully imagine or to expect the full implications or usefulness of the results, although significant spin-offs may indeed result later. Mathematician Hardy, talking about his work, stated that 'I have never done anything useful, no discovery of mine has made, or is likely to make directly or indirectly, for good or ill, the least difference to the amenity of the world. Judged by all practical standards, the value of my mathematical life is nil; and outside mathematics, it is trivial anyhow. I have just one chance of escaping a verdict of complete triviality, that I may be judged to have created something is undeniable; the question is about its value'.

Although science is pursued by all good scientists without any expectations, we cannot forget the immense benefits that we have obtained through science. It is said that when Gladstone, the Chancellor of the Exchequer, asked Faraday about the practical worth of electricity, Faraday seems to have replied, 'One day, Sir, you may tax it'. As scientists, we should take care not to undermine the so-called applied science. Applied science can be as challenging as pure science and, what is more, it is somewhat more difficult. Scientists indulging in fundamental research tend to have a poor opinion of those who do applied science and this attitude has not helped anybody. Science requires many types of workers. To quote Peter Medawar, 'Among scientists are collectors, classifiers and compulsive tidiers-up; many are detectives by temperament and many are explorers; some are artists and others are artisans. There are poet-scientists and philosopher-scientists and even a few mystics'. But one of the main features that characterizes modern scientific research is that 'it is the art of the soluble', unlike politics which is the 'art of the possible'. 'Both are immensely practical-minded affairs. Good scientists study the most important problems they think they can solve'. And to solve real problems of mankind, 'we need to do science and more science' as Raman said. Nehru had the firm belief that 'it is science alone that can solve the problems of hunger and poverty, insanitation and illiteracy, of superstition and deadening custom and tradition, of vast resources running to waste, of a rich country inhabited by starving people.... The future belongs to science and to those who make friends with science'.

Science has been a great instrument of social change, all the greater because its object is not change but knowledge. Its silent appropriation of this important function, amid the din of political and religious strife, is the most vital of all the revolutions associated with the development of modern civilization.

Support for science

Although the tendency in the advanced countries is to support science through sources other than those of Government, there has been no doubt that even in these countries state intervention has been essential to ensure continued, balanced support and to accrue the benefits of the results of scientific work for national development. In countries like India, there is no source other than the Government for the support of science. Unfortunately, what has happened is that there is little appreciation in many quarters regarding the responsibility of the Government in actually ensuring that science is well supported and that the state of health of science is satisfactory. This is because not only some of the ministers, planners and administrators at the highest level are ignorant of science but the same defect runs through almost all the departments of the civil service. It is said to be nearly universal in the House of Commons in Britain, and is shared by the general public, including a large proportion of those engaged in industrial and commercial enterprise. This ignorance is affecting science badly, especially in countries such as India. I am reminded of the story narrated by Al-Biruni. Once a sage was asked why scholars and scientists always flocked to the doors of the rich, while the rich are not inclined to call at the doors of the scientists. The sage answered, 'The scholars are well aware of the use of money but the rich are ignorant of the nobility of science'.

Frederick Dainton wrote sometime ago, 'Accountants and second rate business school jargon are in the ascendent. Costs, which rise rapidly, and are easily ascertained and comprehensible, now weigh more heavily in the scales than the unquantifiable and unpredictable values, and future material progress. Perhaps, science will only regain its lost primacy as people and Government begin to recognise that sound scientific work is the only secure basis for the construction of policies and to ensure the survival of Mankind without irreversible damage to Planet Earth'. I feel that the attitude of many politicians, planners and administrators has not altogether been healthy and has not been based on sound judgement. I recall the statement made by John Kenneth Galbraith regarding Mrs Thatcher's policies in Britain. 'She is the reflection of comfortable middle class values that do not take seriously the continuing unemployment. What I particularly regret is that she does not take seriously, the intellectual decline. Having given up the empire and the mass production of industrial goods, Britain's future lay in its scientific and artistic pre-eminence. Mrs Thatcher will long be remembered for the damage she has done'. I believe that there is a lesson in this statement.

Experimental science

I am a physical scientist. In particular, I am a physical chemist or a chemical physicist doing experiments. I believe that doing experiments is the main purpose of science. In science we have theories and models. As Kapitza said, 'theory is a good thing, but a good experiment lasts for ever'. Leonardo da Vinci seems to have said that 'there is no higher or lower knowledge but only one flowing out of experimentation'. According to Eigen, 'A theory has only the alternative of being right or wrong; a model has a third possibility. It may be right but irrelevant'. Theoretical research without relation to experiment tends to become sterile. I like what Faraday said about experiments. 'Nothing is too wonderful to be true if it be consistent with the laws of Nature and in such things as these, experiment is the best test of such consistency.' Michael Faraday was an experimentalist par excellence and probably the greatest experimentalist of all time.

While it is pointless to discuss the relative importance of experimental and theoretical research, there seems little doubt that when experimental research dies, science also dies in that country. This should be a matter of serious concern to us, since there are very few people doing good experimental work in India, at least in physical sciences and related areas.

Reflections

Science is an essential part of my life. Without science I am unable to live. When I started working in science four decades ago, I did not know that this would happen. I took up science as a profession and as I gradually went on working, I got addicted to it. Soon I found that science was not only running my life but was dictating its terms. There was no way I could plan my life without science. Every aspect of my living is linked to what I do in my laboratory. If my work in the laboratory gets affected by what I do in other spheres of life, then I stop indulging in those activities. I have found that doing science alone is not as interesting as doing science with young people. Being a teacher has therefore been extremely important to me. Being a teacher alone without doing research has no meaning, either. Kalidasa wrote many centuries ago:

> If a professor thinks what matters most, is
> to have gained an academic post, where he
> can earn a livelihood, and then neglect
> research, let controversy rest, he is but a
> petty tradesman at the best, selling retail the
> work of other men.

Those of us who pursue science by working with young people cannot forget our prime responsibility in bringing new generations of scientists. We have to give everything we have to the young and guide them in a manner that they are able to blossom as scientists, fully exploiting their native capabilities. It is difficult to convey the excitement in science unless one has experienced it, 'the dramatic feeling of sudden enlightenment that floods the mind when the right idea finally clinches into space'. Yet, by the way one lives and by the way one works, it is possible that one may be able to communicate the right message to the young. I have found that my Indian background is specially helpful in developing a less selfish and a more purposeful approach to living.

A great diversion to many a good scientist has been to get involved in administration. Unfortunately, it is also a necessary and unavoidable evil. I realize that one cannot possibly be like Michael Faraday. Faraday, when approached to take the Presidentship of the Royal Society, apparently said, 'I must remain plain Michael Faraday to the last; and let me now tell you that if I accepted the honour which the Royal Society desires to confer upon me, I would not answer for the integrity of my intellect for a single year'. This is too much to ask of ordinary mortals like myself, yet there is no doubt that those of us who have done even a little bit of administration have lost valuable time for research and those of us who do research beyond all other commitments, often find that we occasionally miss something crucial and important because of our pre-occupation with administration. Administrative chores and worries frequently come in the way of the crucial minutes of unperturbed meditation. I sincerely advise all those aspiring young scientists who really want to make a mark in science, never to touch any form of administration unless it is just to govern a small research group.

Many scientists worry about status. I have heard people say that an Inspector of Police in the locality has greater respect of the community than a scientist. I do not know what kind of respect they have in mind. I believe that this is not the kind of thing one should worry about. Intellectuals, who own and create knowledge, are the most important component of society and this determines their status and nothing else. We cannot possibly compare the status of a good scientist with that of an Inspector of Police or an administrator in Government. I think that a good scientist does get respect where he belongs, by the international and national scientific community. I also feel that salary and status are irrelevant parameters in science; I have always felt that I have been an overpaid research scholar doing what I like most. It so happens that scientists themselves treat administrators with undue respect and prefer politicians to inaugurate symposia. It is necessary for us to have a good sense of humour. Some of us suffer from a persecution complex, possibly with justification. A man does not attain the status of Galileo because he is persecuted. He must also be a good scientist.

Some scientists are conservative about publishing while others are prolific. While there is no simple rule about how much to publish or not to publish, there is no denying the fact that, as Benjamin Franklin as well as Faraday put it, the main activity of scientists is to 'work, finish and publish'. I am not ashamed of publishing. I have known great scientists who have published several hundred papers. Both Faraday and Raman published over 450 papers. I have also known extraordinary men who have published very few papers. I have known theoreticians, like Nevill Mott, who have published consistently and constantly for several decades. The real problem with us in India, however, is that a large number of us publish a large number of papers which are irrelevant. There are very few good papers coming from India especially in experimental physical sciences, and this is harmful. A small number of good papers coming from a large country like India has little visibility.

Research facilities are poor in most institutions in India. Blackett was of the view that 'A first rate laboratory is one where even mediocre scientists produce outstanding work'. This is probably what happens when some of us go to advanced countries. We must have at least some good laboratories where we can perform and where most of us can do our best to produce results that are worthwhile. While facilities and support are declining, science itself is galloping. The pace at which science is moving is so great indeed that the frontiers of today often become the centres tomorrow and this happens all the time. The present-day rate of progress in science allows no respite and in most areas competition is intense. To be counted, one has to fearlessly take up difficult, contemporary problems. Safe problems are easier to work on, but get little notice.

In spite of the many limitations and difficulties, there are still some of us who want to do science. It is useful, therefore, to recall the lives and work of great men who accomplished much against all odds and would not give up easily. We should not be unnecessarily worried about public opinion. After all, public opinion is not discriminating and is likely to interpret the demand of scientists as meaning that science can be had for money. But science is not for sale.

Freedom may be the first born daughter of science. Unfortunately, however, science seems to have lost her freedom in recent years. 'It has become a productive force' as Kapitza said. 'It has even become rich but it has become enslaved' I am not sure whether men like Faraday, Rutherford and Raman would be happy to do science under the present circumstances.

Samuel Coleridge wrote long ago, 'the first man of science was he who looked into the thing, not to learn whether it furnished him with food, or shelter, or weapons, or tools, or armaments or playwiths but who sought to know it for the gratification of knowing'. I believe that this spirit is still prevalent amongst some of us. Raman said that 'the pursuit of science derives its motive power from what is essentially a creative urge. The painter, the sculptor, the architect and the poet, each in his own way, derives his inspiration from Nature and seeks to represent her through his chosen medium.... The man of science, like the exponents of art, subjects himself to a rigorous discipline, ... Science is a fusion of man's aesthetic and intellectual functions devoted to the representation of Nature'.

I have always felt a sense of admiration and awe by the way Ramanujan worked. He had no need of university education. With all his problems, he went on doing mathematics, writing page after page of equations in his notebooks. What was the urge that made him do that? It was not money. His health was terrible. He had no real patrons until he proved himself and even those patrons probably learnt more from him than he from them. How could he do so much in such a short life amidst suffering and sacrifice?

One cannot help but marvel at the way Michael Faraday carried out experiments. Faraday who hardly had any formal education was a genius propelled by an urge to explore. He was painstaking and dedicated and was a storehouse of intellectual energy. He just would not stop. If one were to look at Faraday's contributions by today's standards, he could have easily received five or six Nobel prizes. No one really guided him or motivated him. It was purely the case of the inner urge inspiring him to carry out experiment after experiment. In the year 1833 alone, Faraday made seven or eight major discoveries. Interestingly, he recorded an important discovery on 24 December and another on 26 December; there is no record for 25 December since it was Christmas Day.

I have been equally baffled by the artistic productivity of Michael Angelo who took five years or more to produce his masterpieces on the extensive ceiling of the Sistine chapel. He painted alone without assistance, day and night. He used to paint hours on end, often in the dark with a candle on his forehead and another on his shoulder, with his head bent backwards to face the ceiling. At the completion of this monumental work, he was temporarily deformed and could not bend his neck. He could not walk because his toe nails had overgrown in the shoes which he had not removed in years.

In my years of working as a scientist, I have found science to be a demanding master, not easy to satisfy. Only those who completely submit to it seem to reap the benefits in terms of intellectual excitement and satisfaction. Those who treat science as a pastime or a hobby, may get little rewards, but nothing else. For a real scientist, all days are working days. For someone mentally absorbed in his work, there are no working hours. Wherever one is, and whatever one is doing, one is always under the effect of the undercurrent of one's scientific pursuits, consciously or unconsciously. When such a thing happens, the need for an external stimulus to pursue science disappears. It is only then that anxiety about recognition and rewards also disappears. This mental state is necessary for philosophically well adjusted living. The effort to attain this state is difficult and may demand considerable personal sacrifice, often in terms of social life; then, it is worth it.

What is also wonderful is that there is no limit to scientific pursuits. This limitless world scientists belong to, makes life worth living and more challenging. I have always been and continue to be thrilled by the way research areas develop as one pursues ideas. 'Great Oaks from little acorns.' As Herbert Brown has written, 'What starts off as a mere grain of pollen develops into an acorn. The acorn then grows into an Oak tree. The Oak tree develops into a forest. We then begin to see the outlines of a whole new continent'. There are undoubtedly many such continents lying undiscovered around us. Much of the life of scientists is spent in search of the grain of pollen or working in a forest. Happy are those who witness the growth of a pollen grain into a continent. This can happen by chance, but chance only favours the prepared mind; it happens mainly due to persistent effort. I have been making all effort possible in the last four decades to seek happiness through scientific explorations, and more so, by keeping company of those who are similarly occupied. The main thing that has happened is that doing science has become a way of life. I am indeed grateful for this blessing.

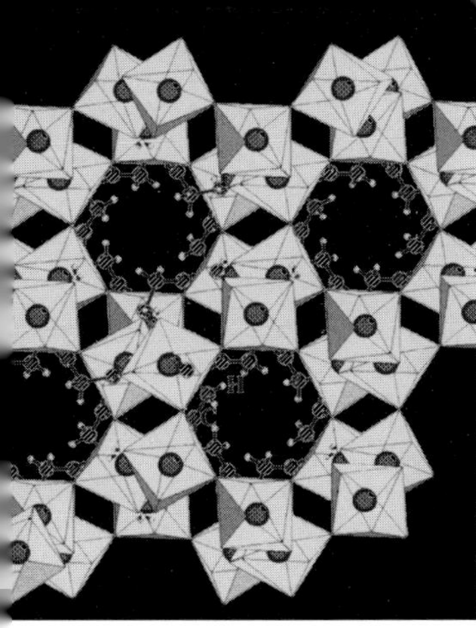

Appendices

1. *Chemistry of Materials: A Letter to a Young Friend*

2. *Science and Spirituality: Personal Reflections*

3. *Science in the Future of India*

4. *Scientific Collaborators*

Chemistry of Materials: A Letter to a Young Friend

C N R Rao

My dear young friend

I am really glad to have this opportunity to write to you, specially about a subject in which I have worked for half a century. When I was your age, if somebody had told me that I would be working in chemistry of materials most of my life, I would not have believed it. At that time, chemistry of materials meant studying something about cement, steel, sand and asbestos. It was indeed dull. I never did well in school and college exams on questions related to this subject. Much later in my life, I got greatly interested in the subject for various reasons. First, in my study of chemistry, I was influenced by Linus Pauling who is my academic grandfather. His book titled 'Nature of the Chemical Bond' which I read when I was young made a great impression. It taught me how the structure of molecules and materials is an extremely important aspect of chemistry, and how the understanding of structure and bonding provides a basis to understand chemistry and to do new chemistry. It is because of this terrific inspiration that I started studying chemistry. It was clear at the end of my undergraduate career that I wanted to be a chemist.

It was at Purdue University when I was finishing my work for the PhD degree (around 1957), that a friend of mine came to me with a sample of spectroscopically pure TiO_2. "What is the structure of this?", he asked me. He thought that it had the rutile structure. "Why don't we determine the structure?", I asked. He wanted me to do this for him. I took the X-ray diffraction pattern with the diffractometer in the physics department and found that it had the anatase structure which is the other form of TiO_2. We spent some time to investigate when

*** Reprinted from Resonance, May 2009, p. 508–517**

and how pure anatase transforms to the rutile structure. This is one of the early papers on solids that I published. My interest in the subject continued as I started looking at papers of a similar nature and started publishing little notes here and there. Then I thought that I should study the chemistry of solids more, to understand their transformations, properties and other aspects. Thus, my early work as an independent researcher related to such things as structural transformations in oxides, halides, etc., decomposition and oxidation of solids, and properties of solids such as electrical conductivity as a function of composition. As time went on, my interest in the subject increased because of the way the subject transformed.

The story of Chemistry of Materials is quite interesting. It is not as if chemists straight away started working on the subject as we understand it today. It has taken quite some time for the subject to develop to the present stage. In the 1950s, very few chemists were working in an area which was known as Solid State Chemistry. Most of the solids were inorganic. The prime concern at that time was to understand what we meant by stoichiometry. The problem was the following. There are compounds like NaCl which were easy to understand, with one Cl for every Na. But, there are also other compounds which have unusual compositions such as Pr_6O_{11} and Ti_4O_7. If we assign a charge or oxidation state of –2 to oxygen, Pr and Ti will end up with non-integral charges or oxidation states. How does one rationalize compositions of this kind? This was a big puzzle for chemists. There were people who thought that compounds were not necessarily stoichiometric. Then, how does one explain the absence of simple stoichiometry? It was only in the late 1960s that we understood that such unusual compositions arose because of new types of structural manifestations in solids. They were not due to the presence of defects like vacancies. The subject of solid state chemistry gained maturity by the 1960s. You will be surprised to know that there were no journals to publish research papers in solid state chemistry at that time. It was only in 1968 that the first journal devoted to solid state chemistry was established. (Today, we have several journals dealing with chemistry of materials). As time went on, people worked on many aspects of solid materials particularly metal oxides of various kinds.

Then started intense activity in other classes of materials, including organics.

It is no surprise that in the last few years, the two main streams of chemistry are directed towards biology and advanced materials. It is also not difficult to understand why chemists are getting more and more interested in materials. After all, the quality of life of human beings is directly related to the availability of materials of various kinds. When I say materials, these are not just common materials such as wood, coal, cement and steel, but the vast variety of materials required for electronics, computers, transportation, energy, etc. Materials chemistry encompasses all kinds of materials, organic, inorganic and biological. Materials chemistry deals with materials of all kinds and in all states of matter (Figure 1). They can be glassy, they can be crystalline, they can be liquids, they can be soft solids and they can be hard. The materials may have one of many properties, be they electronic, magnetic, dielectric, mechanical, adsorptive or catalytic. Tailor-making materials with desired or controllable properties is one of the prime objectives of materials chemistry today.

I mentioned that metal oxides have constituted a major area of work in the chemistry of materials. You may ask why metal oxides? Metal oxides form a class with the widest range of properties. There

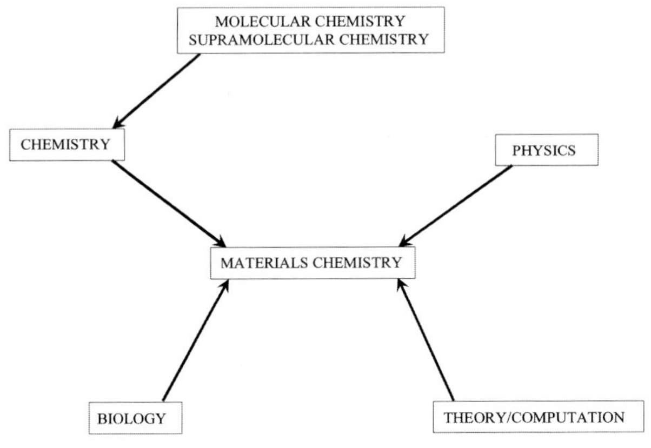

Figure 1. What constitutes materials chemistry.

are oxides which conduct like metals, and there are oxides which are superconductors. There are also oxides which are the best insulators or dielectrics. Some oxides are magnetic. Metal oxides, therefore, have occupied the attention of solid state chemists in a big way. Chemists have synthesized oxides of various structures and properties. I remember in the 1960s, an oxide of rhenium, ReO_3, was made. ReO_3 looks like copper and conducts electricity like copper. It made big news. Soon, many other materials of this kind were discovered. There are also metal oxides which are metallic like copper at certain temperatures and become insulators like wood at lower temperatures. Such properties aroused much interest in me and others.

The culmination of the interest of chemists in metal oxides was when high-temperature superconductivity was discovered. This was in late 1986. Till then, the highest superconducting transition temperature known to us was 23 Kelvin. In December 1986, it was announced that there was an oxide containing copper which became superconducting around 35 Kelvin. It broke the 23 Kelvin barrier and created a big sensation. People all over the world started working on superconductors. Chemists contributed in a big way by making a variety of oxides with superconducting properties. The first attempt in 1987 was to make an oxide which became superconducting at or above liquid nitrogen temperatures (i.e., above 77K). Soon, $YBa_2Cu_3O_7$ with a transition temperature of around 90K was discovered. I am glad to have been part of that effort. I have not seen in my life any discovery that has caused greater sensation than high-temperature superconductivity. Thousands of papers have been published on the subject and it has consumed nearly two hundred thousand pages of journal space. The highest transition temperature that we have attained for superconductivity to date is around 135 Kelvin. It is an oxide containing copper, mercury, and barium. If one applies high pressure on it, it becomes superconducting around 165 Kelvin. I am told that somewhere in the world, room temperature is around 160 or 170 K. If so, this material will be a room-temperature superconductor at that place. Ordinarily, in places like where you and I live, we still do not have a room-temperature superconductor. It will be wonderful if somebody can discover superconductivity at room temperature. It would change the world. (Can you guess why?).

Inorganic	**Organic**
Semiconductors	Semiconductors
Metals	Metals
Superconductors	Superconductors
Magnets	Magnets
Ferroelectrics	Ferroelectrics
Non-linear optical materials	Non-linear optical materials

Table 1. Materials can be inorganic or organic.

Although I have mentioned only metal oxides till now, there are many other classes of inorganic solids such as nitrides and sulphides which exhibit fascinating properties and have immense applications. For example, GaN on excitation gives blue light. If one mixes yellow light with it using a phosphor, we get white light. This is the methodology used for solid-state lighting. In the last two to three decades, organic materials have gained great importance. People have generated organic compounds which conduct like metals. Some also exhibit superconductivity. There are organic magnets. Thus, there are organic materials which exhibit properties similar to inorganic materials (Table 1). Studies of solid state reactions involving organics and the development of the principles of crystal engineering have given rise to a new facet of materials chemistry. It will be of great interest if one can work on materials which have both organic and inorganic parts, what you might call hybrid materials. Properties and phenomena associated with hybrid materials would be most exciting. In the meantime, let me mention something about polymers.

Polymers of various kinds are known. There are structural polymers with high mechanical strength. They are being used everywhere and they can replace steel and other construction materials. Polymers which have unusual electronic and optical properties have been made. There are polymers which give out blue light on excitation, somewhat like gallium nitride. Polymers have been used to make transistors. Novel polymeric devices such as solar photovoltaic cells and transistors have also been made. Polymer electronics or plastic electronics has, in fact, become an important area in the chemistry of materials.

Materials chemistry today makes use of all the advances of chemistry, and tools of chemistry. It applies all available synthetic, structural and theoretical tools. It makes use of novel strategies such as supramolecular organization and self assembly. It uses biological principles and processes. Let me tell you something that may interest you. As you know, Nature makes hundreds of tons of silica every year. It would be nice to know how Nature makes so much silica under neutral pH. People have found that Nature uses a specific protein to hydrolyze silicon compounds to produce silica. Similarly, people have been interested to know how sea shells are made. Sea shells are composed of calcium carbonate. Nature finds the right kind of proteins to build the beautiful shell structures by arranging bricks of calcium carbonate. Some of these shells make use of calcium carbonate, not in the normal calcite structure, but in the metastable aragonite structure.

As you can guess by now, the subject of materials chemistry is becoming more and more interdisciplinary. To really contribute to it, one has to know a fair amount of chemistry and physics, and where necessary some biology. Those who do theoretical and computational work will also require some expertise in mathematics and with computers. This is what makes the subject attractive to me. Important contributions in materials have actually come from such an interdisciplinary approach. I mentioned about superconductivity earlier. It was considered to be a topic in physics, but today after the discovery of higher-temperature superconductivity, it has become part of chemistry as well. In fact, chemistry got its due recognition in the area of materials after the advent of high-temperature superconductivity. In the last few years, many unusual compounds of iron and cobalt have been made by chemists which are found to show superconducting properties.

An important point to remember is that science progresses continuously. This is very much true of materials chemistry. Periodically, there are new discoveries of materials or methods that revolutionize the subject, and change the very direction of the subject (Table 2). Let me give a couple of examples. If you remember, in the year 1985, fullerenes C_{60} and C_{70} were discovered. They were actually made in the laboratory in the year 1990. Soon after it became possible to make fullerenes in the laboratory, a large number of chemists made use of fullerenes in

1985	—	Fullerenes
1986	—	High-temperature superconductivity
1991	—	Carbon nanotubes
1992	—	Mesoporous solids
1993	—	Colossal magnetoresistance in rare earth manganites
2005	—	Graphene

Table 2. Some recent discoveries.

materials design. Fullerenes are being used in a variety of situations. An extended form of fullerene is the carbon nanotube. Carbon nanotubes were discovered in the year 1990 and they have become important materials because of their high mechanical strength and other important properties.

There is another phenomenon that I would like to mention. This phenomenon is colossal magnetoresistance. Let me explain what magnetoresistance is. If you measure the electrical resistance of a material at some temperature, you get some value. Then you apply a magnetic field. The magnetic field may decrease or increase the resistance of the material. This increase or decrease in resistance due to a magnetic field is called magnetoresistance. If a material exhibits large magnetoresistance at ordinary temperatures, on applying a small magnetic field, it will have tremendous applications in computers and other areas. One of the major discoveries in 1993 was the observation of very high or colossal magnetoresistance in certain oxides of manganese. This subject has given rise to considerable amount of research. You can again see how oxides are important.

Another class of oxide materials called multiferroics is causing great interest in academia as well as in industry. What are multiferroics, you may ask. As you know, some materials are magnetic while some others may show a maximum in dielectric constant at some temperature. Ferromagnetic materials show hysteresis when magnetization is plotted against magnetic field. Some of the dielectrics (ferroelectrics) exhibit spontaneous polarization on application of an electric field and the polarization shows hysteresis with respect to the electric field. A material possessing such magnetic and dielectric properties simultaneously is called a multiferroic. What would be interesting is if one can find a material which shows electric polarization on the application of a magnetic field, or show magnetism on the application of an electric field. Such a material would be even more interesting. There is considerable activity all over the world to discover new materials which exhibit such properties.

Porous solids are of great use for adsorbing gases and vapors as well as in catalysis. Don't forget that zeolites are used commonly in various catalytic applications. Zeolites have channels where only molecules of a particular size (and shape) can get in. Solids with micro-, nano- and meso-pores have been made in the last two decades or so. In recent years, compounds with structures similar to zeolites have been made based on metal phosphates, carboxylates and sulfates. Metal carboxylates are specially interesting since they contain both organic and inorganic components. They are hybrid materials. There is much to be done on hybrid materials wherein we use the functionality and flexibility provided by the organic part along with the rigidity of the inorganic units (see Figure 2 for example). One of the useful properties of some of these compounds is their ability to adsorb hydrogen.

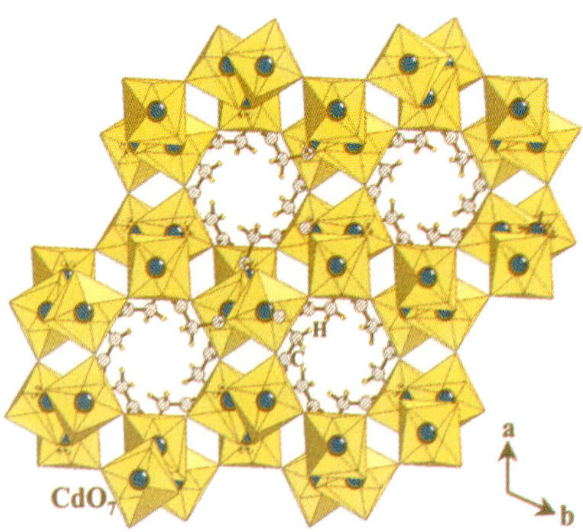

Figure 2. A simple three-dimensional hybrid metal carboxylate.

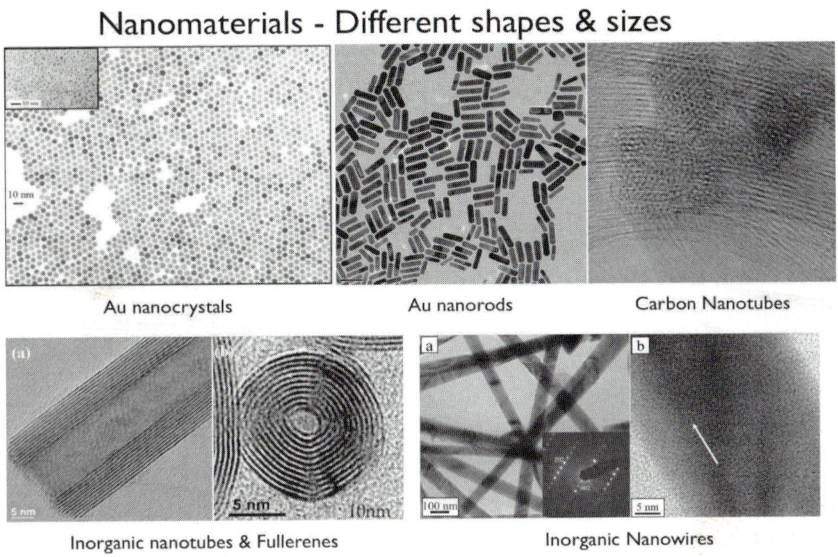

Figure 3. Nanomaterials.

In the last few years, nanoscience has become a favourite subject. (Remember a nanometer is 10^{-9} meter or 10 Angstroms). In nanoscience, a high percentage of the work relates to chemistry. Chemistry of nanomaterials which includes synthesis, self-assembly and other aspects is an important aspect of nanoscience. In the last few years, people have made nanomaterials with different dimensionalities – nanorods, nanowires, nano-tubes, nanodots and nanoparticles of a large number of inorganic materials (Figure 3). Chemical methods have become essential to make most of the nanomaterials and it appears that we can make almost any material in nanoform by employing an appropriate method. An unique feature of nanomaterials is that size determines their properties. In the nanoregime, new properties manifest themselves due to quantum effects. Nanomaterials have been characterized in various ways, specially by using a variety of microscopes. Many applications of nanomaterials have become possible already. Nanoelectronics still poses many probelms. One challenge would be to incorporate nanoscience with molecular electronics for possible applications.

There are many more materials of equal interest both from the inorganic and organic chemical world. There are several fascinating biological materials. Chemists have tried to mimic biology or find substitutes for natural materials such as bone and the spinal cord. There are several other aspects in biology that have directly influenced the practice of materials chemistry. In addition to solid materials, there is much activity in areas dealing with gels, liquid crystals and such soft materials. Chemistry of amorphous or glassy materials is a vast subject by itself.

I trust that I have been able to give you a flavor of materials chemistry. To be good at it and do something useful, is not only exciting but also demands various types of abilities as I had mentioned earlier. What is really nice is the following. If creativity and innovation are important aspects of science, and necessary needs of society, then materials chemistry has a special place since it requires extraordinary creativity and innovation. One is always creating new materials that didn't exist before. One is finding new properties not looked at before. This fascinates me. Whatever I have tried to do involves creating new materials and looking at novel phenomena. I hope that the subject interests you and other young people. A capable young scientist working in this area can contribute to many aspects which directly deal with the quality of life. Coming to the quality of life, let me say one or two words. Look at the present-day problems of mankind. The pressing problems today have to do with energy, climate and so on. In some countries, safe drinking water is not available. The only way we can solve the problem of energy is by finding the right kinds of materials for beneficiating solar energy and storing hydrogen as well as by designing better batteries and fuel cells. Let us remember that a compound that can store hydrogen upto 6 wt% can be used in a hydrogen fuel cell to run automobiles. All this can happen only with the contribution of materials chemists. Even producing safe drinking water has a component of materials chemistry. One may like to use certain nanomaterials and new kinds of filters for the purpose. I believe, therefore, that there is unlimited scope and unlimited opportunities to contribute to science and serve mankind through materials chemistry.

I can go on and on, but I must stop. You must realize that there are so many classes of materials and so many varied aspects related to them. I cannot possibly mention it all in a little letter, but I hope that this letter has been sufficient to tell you why people do chemistry of materials and how interesting it can be because of its interdisciplinary and futuristic nature. I do hope to see you one of these days. If you ever decide to work in chemistry of materials, do write to me. You are most welcome to spend some time in my laboratory.

With affectionate good wishes,

(C.N.R. Rao)
National Research Professor and
Linus Pauling Research Professor

Jawaharlal Nehru Centre for Advanced
Scientific Research
Jakkur P.O., Bangalore 560 064
India
Phone: + 91 80 23653075/22082761
Fax: + 91 80 22082760
E.mail: cnrrao@jncasr.ac.in
Webpage: http://www.jncasr.ac.in/cnrrao

P.S : *I forgot one thing. Theory and computation have succeeded immensely in the last decade or so in predicting the structure and properties of molecules and materials. So, if one does not like to make or measure, one can calculate or simulate.*

If you and your friends want to read a book that may give an introduction to materials chemistry, I suggest, 'New Directions in Solid State Chemistry' by C N R Rao and J Gopalakrishnan, Cambridge University Press (available as a paper back)

"The devil may write chemical text books because every few years, the whole thing changes."

– *Berzelius* (1779–1849)

*Science and Spirituality: Personal Reflections**

Prof. C N R Rao

Science and spirituality have many common attributes. Scientists seek to understand the universe: the matter, the forces, and the phenomena governing it. In their quest for understanding and desire to attain the highest form of knowledge, scientists also seek perfection. Spirituality is no different. Those who are on the spiritual path seek perfection and a complete knowledge of the ultimate reality. Both science and spirituality seek to understand the secret of life and being. A good scientist has to keep climbing the limitless ladder of excellence, since there is no real end to the pursuit of or to the effort to attain perfection, for no one ever reaches the perfect state. A seeker of spirituality similarly must climb the limitless ladder of excellence, because humans can never reach the state of perfection. In defining the perfect one, I can only think of a God who is omniscient and omnipresent, a God who is expansive, kind, and forgiving. We are all the children of the same God, possessing the same atoms created at the beginning of the universe. We are his images. According to Madhvacharya, one of India's religious seers and philosophers, 'This world is the only reality, and it is not a myth. He (God) is supreme, and we are all his images. How good the image is depends on one's qualities.'

Albert Einstein said that without a personal philosophy and a religious bent of mind, it is impossible to live. Only insecure and shallow persons question this. In this context, I find it important to be honest in expressing one's inner feelings. Many people have asked me, 'Do you believe in God?' I am not shy to say that I believe in God. I do not however, believe in superstitions or omens. I do not have my horoscope (neither did my father), but I enjoy going to temples. In fact, I visit the sacred shrines of all religions. I have spent hours in quiet

*** Reprinted from Prabuddha Bharata (2008), P. 63–64**

contemplation in many temples of India and in St Peter's Basilica in the Vatican. I have prayed at the Holy Wall and the shrine above it in Jerusalem.

The uncertainties in this world, the way things happen and the way nature works, demand a certain attitude. I think there are things which are beyond simple rationale. Everything cannot be explained away in simplistic terms. This is true of love, beauty, and music. We should have the humility to say that we do not understand everything. Such humility is essential of true scholars. It is actually not necessary to be able to explain every emotion and aesthetic experience.

I believe in a God who is all-powerful. His omnipotence and omnipresence make it impossible to describe him in one particular way. For this reason, people worship him in different forms. I am not surprised by this. Maybe that is why we have so many gods and goddesses in India. Each time we probably worship one aspect of God. But one thing is definite. It is essential to pray in order to have a definite way of life, to have guidelines for ourselves, and to live harmoniously in this world. In this context the theory of karma is useful; and believing in the next life is helpful to be a good person in this life.

My faith in God has helped me in all my endeavours. Where does a person like me get the moral courage to face the world with all its problems? How is it that I try to accomplish and practise science at the highest possible level, working under conditions which are not necessarily the best and in an atmosphere which may not be the most conducive? Does this require some help? Yes. I believe that the moral courage that I have needed to face various problems in my life has been the gift of God. There are many things that bother us all—for example, when family problems become difficult, when we are left out, when we are insulted, when our own colleagues pull us down, when we do not get necessary encouragement, and when our own students or younger people behave badly. These problems are common everywhere, but particularly so in India. How do we get the forbearance to keep on smiling? How do we keep our aim high and retain our ideals in spite of everything else that pulls us down? This can be done only with the help of God almighty.

Genuine scientists as well as seekers of spirituality have to be single-minded. In fact, single-mindedness or perseverance is the single most important quality for a scientist to accomplish something worthwhile in India. This reminds me of a story that my mother, who was also my first teacher, used to tell me. She used to say that Saraswati, the goddess of learning, is extremely difficult to please, while it is much easier to please Lakshmi, the goddess of wealth. If you pray to her a little, Lakshmi will bless you with some money. However, it is very difficult to get the blessings of Saraswati. In order to attain knowledge and wisdom, one has to surrender to her absolutely. It is only then that Saraswati may look at you. If one has to become a great scholar or research worker, complete surrender to Saraswati is an essential pre-requisite. When eventually Saraswati blesses one, all other rewards will also come to the person. I think it is the lack of this single-mindedness that has affected the state of development of science and scholarship in this country. It is best to gain everything in life through knowledge.

However single-minded one might be, one cannot help getting doubts about one's chosen path. Doubts about oneself and about various other matters arising from everyday life can be unnerving and debilitating. When such things happen, I remember the Bhagavadgita and what Mahatma Gandhi said about it: 'When disappointments stare me in the face, and when I see not one ray of light on the horizon I turn to the Bhagavadgita, and find a verse to comfort me; and I immediately begin to smile.'

To my knowledge, the foremost person in India who has served God through service to humanity in an extraordinary manner is Amma (Mata Amritamayeeji)

*Science in the Future of India**

C. N. R. Rao

India has voted for Science. In May, half a billion people cast their ballots, and they decisively favored spurring the development of the world's second most populous nation. The reelected Prime Minister Manmohan Singh and his new coalition government have made a commitment to reduce poverty and disease, create employment, and stimulate rural and industrial development. Attaining these goals will require substantial new investments in science and technology (S&T) plus much greater investments in human capital.

Since achieving freedom in 1947, India has established many institutions devoted to science and higher education. Most notably, five Indian Institutes of Technology (IITs) were established between 1951 and 1963, and by 2008 there were 13 IITs: national degree-granting institutions devoted to the training of high-quality engineers and scientists. Despite the gap in infrastructure between advanced countries and India, there have been critical successes in areas such as space, atomic energy, and agriculture. In fundamental research too, India has made progress. Because of the innumerable demands on the economy, however, the higher-education sector has not received adequate support. Part of the reason for the decline in India's university science education system in the past decades has been the preferential funding for R&D activities in national research laboratories.

Prime Minister Singh has recently announced an increase in government investment in S&T from the present 1% of gross domestic product (GDP) to 2% of GDP over the next year or two, an increase of unprecedented magnitude. The contribution of industry has also increased significantly in the past few years, now amounting to approximately 20% of the nation's total investment in science R&D.

***Reprinted from Science, 325 (2009) 126 (Editorial)**

And the government has begun appropriate administrative reforms as well. For example, two new government departments dealing with Earth system science and health research have been created. In addition, the Indian parliament has approved creating a National Science and Engineering Research Board, an entity somewhat similar to the U.S. National Science Foundation (NSF), that will be responsible for funding scientific research. It will provide competitive grants and establish new facilities in priority areas. Like NSF, the board will also produce annual "science indicators": detailed analyses for measuring progress in S&T from year to year.

This is all good news. But the human resources essential for supporting an expanded S&T agenda are lacking. Young graduates today are readily attracted to professions other than those related to science and engineering; thus, banking, business, and information technology have become immensely popular. India must now focus on creating a large body of outstanding young people interested in taking up professions in science and engineering. To improve the quality of the university education system, new support is being provided. For example, five new Indian Institutes of Science Education and Research have been established in the past 3 years. Admitting undergraduates on the basis of competitive examinations (as do the IITs), these new national institutes will encourage bright young students to pursue science as a career, at both the undergraduate and Ph.D. levels. In addition, to meet the demand for top-class engineering graduates nationally and internationally, the country will increase the total number of IITs to 15.

Sixty percent of the Indian population is below the age of 25, and most reside in villages. This untapped talent represents a great potential asset. Around 600,000 scholarships are now available each year for talented school students from these areas, with an emphasis on those living in poverty. One million science awards are being given to students to promote interest in science, and 10,000 scholarships are available to support students who wish to pursue education beyond high school. In addition, the new government has already initiated important structural reforms in the education sector.

India's citizens have risen to the occasion with their vote. The tasks and challenges for the new government are clear but daunting: It must now satisfy the aspirations of a billion people.

Scientific Collaborators

Post-graduate Research Students

Agarwal, D.C.
Agarwal, U.
Agrawal, V.
Aiyer, H.N.
Arulraj, A.
Anupama, T.R.
Arunarkavalli, T.
Ayi, A.
Ayyappan, S.*
Bahadur, D.
Balasubramanian, A.
Basu, P.K.
Behera, J.N.
Bhaskar, K.R.
Bhat, S.N.
Bhat, S.V.
Biswas, K.
Chakrabarti, A.
Chaturvedi, G.C.
Chitara, B.
Choudhury, A.*
Dan, M.
Das, B.
Deepak, F.L.
Dwivedi, P.C.
Ebenso, E.E.
Eden, C.
Ganapathi, L.
Ganguli, A.K.*
Ganguly, S.*
Gautam, U.
Ghosh, Anupama
Ghosh, Anirban
Ghosh, M.
Ghosh, S.
Gomathi, A.
Gosavi, R.K.
Govindaraj, A.*
Gundiah, G.
Gupta, A.
Harikumar, K.R.
Ita, B.
Jayaram, V.
Jayaraman, K.
Kalaiselavam, M.
Kalyani Kutty
Kamath, P.V.
Kannan, K.R.
Kulkarni, G.U.*
Kundu, A.
Madhusudan, W.H.
Mahesh, R.
Maitra, U.
Manivannan, V.
Matte, H.S.S.R.
Mehrotra, P.N.
Mohan Ram, R.A.
Murthy, A.S.N.*
Murugavel, P.
Murugesan, T.
Nagarajan, R.*
Natarajan, M.
Nath, M.
Neeraj, S.*

---

*also worked as postdoctoral associates

Om Parkash
Panchakarla, L.S.
Parashar, S.
Parthasarathy, R.
Patil, K.C.
Prabhakar, S.
Prabhakaran K.
Pradeep, T.*
Ch. Pulla Rao
Raidongia, K.
Raina, G.
Raju, A.R.*
Rajumon, M.K.
Rama Rao, G.
Ramasesha, S.*
Ramasesha, S.K.
Ramdas, S.*
Ranga Rao G.*
Rangavittal, N.
Rao, A.M.K.
Rao, K.G.
Rao, K.J.*
Rout, C.S.
Sahu, J.R.
Sankar, G.*
Santra, A.K.
Sardar, K.
Sarma, D.D.
Sastry, R.L.N.
Satish Kumar, B.C.
Seikh Motin
Sen, P.
Sen, R.
Serrao, C.R.
Seshadri, R.
Shukla, A.K.*
Singh, S.
Sivashankar, K.
Somasundaram, P.
Srinivasan, A.
Srinivasa Gopalan R.
Subbanna, G.N.
Subba Rao, G.V.*
Subrahmanyam, K.S.
Sudheendhra, L.
Swamy, H.R.
Tantry, K.N.
Thomas, P.J.*
Thirumurugan, A.
Uppal, M.K.
Vanitha, P.V.
Varghese, N.
Varma, V.
Vasudevan, S.
Vaidhyanathan, R.
Venkataraghavan, R.
Vidyasagar, K.
Vinod, C.P.
Vijayakrishnan, V.
Vijayaraghavan, R.
Vijayasarathy, K.
Vivekchand, S.R.C.
Voggu, R.
Yashonath, S.
Yoganarasimhan, S.R.

Other Post-graduate Collaborators

Bouet, L.
Buttrey, D.
Chakraborti, P.
Chandrabhas, N.
Ghosh, A.
Guha, A.

Jadhao, V.
Jagadeesan, D.
Kalyanaraman, V.
Kam, K.
Ch. Laurent
Loehman, R.
Mahendiran, R.
Nanjundaswamy, K.S.
Ramachandran, J.
Ramanan, A.
Singh, K.K.
de Santos (Antonio)
Singh, K.P.
Somasundaram, T.
Sreedhar, K.*
Thornton, G.
Vasan, H.N.
Vasanthacharya, N.Y.

Postdoctoral Associates

Bandekar, N.
Bharati, S.
Bhat, V.
Bhattacharya, R.
Chandrashekhar, G.V.
Chandrashekhar, M.
Chatterji, S.
Desiraju, G.R.
Durrant, M.C.
Eswaramoorthy, M.
Fernández, J.R.
Ganguly, P.
Grantscharova, E.
Gundakaram, R.
Hegde, M.S.
Husseini, J.
Jagannathan, K.
Jagannathan, N.R.
Kudchadkar, S.A.
Kumar, P.
Kumardhas, P.
Late, D.
Nag, A.
Nagaraj, M.
Pedireddi, V.
Prakash, B.
Ponnambalam, V.
Rao, K.P.
Randhawa, H.S.
Reller, A.
Santosh, P.N.
Sarode, P.R.
Schwenzer, B.
Singh, R.S.
Sreekanth, C.S.
Sugandha
Sumathy, R.
Tomar, M.S.
Ulagappan, N.
Viswanathan, B.
Verelst, M.
Yadav, R.T.

Senior Collaborators

Agarwala, U.C.
Ahluwalia, J.C.
Anderson, J.S.
Attfield, P.

Balaram, P.
Balasubramanian, D.
Bando, Y.
Banerjee, D.K.
Bhattacharya, S.
Becker, E.D.
Bhat, S.V.
Bhide, V.G.
Bose, M.
Chakravorty, D.
Chakraverty, B.K.
Chandrabhas, N.
Chandrasekaran, S.
Cheetham, A.K.
Clifford, A.F.
Edwards, P.P.
Ferraro, J.R.
Feuer, H.
Ferey, G.
George, M.V.
Goodenough, J.B.
Gopalakrishnan, J.
Govindaraj, A.
Green, J.C.
Guru Row, T.N.
Honig, J.M.
Jefferson, D.
Jones, W.
Jortner, J.
Kasturi, T.R.
Kern, S.
Krishnan, V.
Kulkarni, G.U.
Krupanidhi, S.B.
Kumar, N.
Mahanty, J.
Maitra, U.
Maji, T.
Mittal, J.P.
Mott, N.F.
Mueller, A.
Murugavel, R.
Nagarajan, R.
Natarajan, S.
Nordblad, P.
Orville Thomas, W.
Pati, S.
Proffen, T.
Ramakrishnan, T.V.
Ramamurthy, U.
Ramasesha, S.
Rao, K.J.
Ratajczak, H.
Raychaudhuri, A.K.
Raveau, B.
Roberts, M.W.
Roesky, H.
Rousset, A.
Sampathkumaran, E.V.
Sanyal, M.
Singru, R.M.
Sood, A.K.
Smith, P.
Srinivasan, R.
Sundaresan, A.
Tanaka, Y.
Tenne, R.
Thomas, J.M.
Venkatesan, K.
Vishwamitra, M.A.
Vogt, T.
Waghmare, U.
Woodward, P.M.
Yashonath, S.

Collaboration with fellow Students (1954–59)

T.S. Chao
A. Czanderna
P.A. Faeth
G.K. Goldman
L.V. Gregor
C.W.W. Hoffman
L. Kaplan
M.J. Lewis
C. Lurie
R. Oftedal
W. Page
C.N. Pillai
T.V. Rajakumar
L. Rocks
J. Ramachandran
V.S. Ramachandran
H. Rubinstein
C. Savides
G.B. Silverman
S. Somasekhara
W. Wahl
E. Williams

The chemists are a strange class of mortals who seek their pleasures among soot and flame, poisons and poverty, yet among all these evils I seem to live so sweetly that may I die if I would change places with the Persian king.

John Joachim Becher (1625–82)

Index